The Laboratory of the Mind

Thought Experiments in the Natural Sciences

Second Edition

James Robert Brown

Routledge
Taylor & Francis Group

NEW YORK AND LONDON

First edition published 1991
by Routledge

This edition first published 2011
by Routledge
270 Madison Ave, New York, NY 10016

Simultaneously published in the UK
by Routledge
2 Park Square, Milton Park, Abingdon, Oxon OX14 4RN

Routledge is an imprint of the Taylor & Francis Group, an informa business

First edition © 1991 James Robert Brown
Second edition © 2011 Taylor & Francis

Typeset in Sabon by Swales & Willis Ltd, Exeter, Devon
Printed and bound in the United States of America on acid-free paper by
Walsworth Publishing Company, Marceline, MO

Library of Congress Cataloging in Publication Data
Brown, James Robert.
    The laboratory of the mind : thought experiments in the natural sciences /
James Robert Brown. — 2nd ed.
    p. cm.
    [etc.]
    1. Physics—Philosophy. 2. Thought experiments. 3. Quantum theory.
    4. Science—Methodology. 5. Rationalism. 6. Knowledge, Theory of.
    7. Philosophical issues in science. I. Title.
    Q175.B7965 2010
    530.01—dc22
    2010001651

ISBN 13: 978-0-415-99652-5 (hbk)
ISBN 13: 978-0-415-99653-2 (pbk)
ISBN 13: 978-0-203-84779-4 (ebk)

Contents

Preface and Acknowledgements
(First Edition)

Like most philosophers, I encountered a bit of rationalism (Plato and Descartes) and a bit of empiricism (Hume) in my first formal introduction to the subject. And like most students of philosophy I found the rationalists endlessly fascinating, but not in the least believable. It seemed obvious to me, as it does to most, that all our knowledge is based upon sensory experience. Then one day I heard about Galileo's thought experiment showing that all bodies must fall at the same rate—I almost fell out of my chair. It was a wonderful intellectual experience. Suddenly, traditional rationalism seemed a live option; perhaps my philosophical heroes—Plato, Descartes, and Leibniz—were on the right track after all.

All of this remained on the back burner until a couple of years ago when I got around to looking at thought experiments in general. I was surprised by two things. First, that there is remarkably little literature on the subject. People often use the expression 'thought experiments', but hardly anyone has thought seriously (or at least written extensively) about them. The second thing I was surprised at was that my old rationalist sentiments stood up; if anything, they have been reinforced by looking at this topic anew. I have long held a Platonistic view of mathematics; I now hold a Platonistic view of physics as well.

In brief, the book is as follows. The first chapter introduces the subject by giving several examples of thought experiments. A multitude of cases is necessary since I have no definition of *thought experiment* to work with; we need a variety of paradigm instances. But this is not the only reason for describing several specimens. They are such a pleasure to contemplate that it's an opportunity not to be missed.

With lots of examples under our belt we can begin to talk about how they work. This task is begun in chapter two, which offers a taxonomy of thought experiments. Some commentators say thought experiments do this or that or some other thing. Actually they do several quite distinct things and chapter two tries to classify their diverse uses.

The Lab

Newton's bucket, Einstein's elevator, Schrödinger's cat—these are some of the best-known examples of thought experiments in the natural sciences. But what function do these experiments perform? Are they really experiments at all? Can they help us gain a greater understanding of the natural world? How is it possible that we can learn new things just by thinking?

In this revised and updated new edition of his classic text *The Laboratory of the Mind*, James Robert Brown continues to defend apriorism in the physical world. This edition features two new chapters, one on "counter thought experiments" and another on the development of inertial motion. With plenty of illustrations and updated coverage of the debate between Platonic rationalism and classic empiricism, this is a lively and engaging contribution to the field of philosophy of science.

James Robert Brown is Professor of Philosophy at the University of Toronto.

Chapter three [now chapter four] is a defence of Platonism in mathematics. It serves as a model of how I'd like theorizing about thought experiments to go. Platonism in the philosophy of mathematics, though a minority view, is eminently respectable, whereas a priorism about the physical world is likely be dismissed out of hand. So the point of the third [fourth] chapter is to carry empiricist-, naturalist-, and physicalist-minded readers as gently as possible into chapter four [five] which contends that we really do have some *a priori* knowledge of nature. Of course, the great bulk of our knowledge must be accounted for along empiricist lines; but there is, I contend, the odd bit that is *a priori* and it comes from thought experiments. Not all thought experiments generate *a priori* knowledge. Only a very select class is capable of doing so. This *a priori* knowledge is gained by a kind of perception of the relevant laws of nature which are, it is argued, interpreted realistically. Just as the mathematical mind can grasp (some) abstract sets, so the scientific mind can grasp (some of) the abstract entities which are the laws of nature.

The next two chapters function as a kind of test of my Platonist outlook, though I hope there is some independent interest in these final chapters as well. A novel interpretation of Einstein is offered in chapter five [seven]. It attempts to make sense of what is commonly thought to be Einstein's 'youthful empiricism' and his 'mature realism' as well as accounting for the role of thought experiments in his scientific work.

Chapter six [eight] surveys some of the philosophical problems of quantum mechanics and some of the interpretations which have been proposed to solve them. Though my discussion of thought experiments has been mainly about the epistemology of science, much ontological machinery has been developed. Laws of nature, understood as real entities in their own right, are now put to work to give an account of quantum phenomena—not just thought experiments about quantum phenomena, but the actual measurement results themselves—which is realistic and does not violate the locality requirements of special relativity.

It is a commonplace at this point in a preface for authors to say that they are less interested in having their own views stand up to close scrutiny than in stimulating interest in their chosen subjects.

Of course, such remarks are largely disingenuous. Nobody wants to be shot down in flames—certainly I don't; yet the chance of this happening is great since the central views put forward here are far removed from mainstream thinking about how either science or nature work. Even if some of my claims are on the right track, the details are bound to be seriously flawed. At best this work is a first attempt at a (modern) rationalist interpretation of science. So it is probably wise for me also to take the disingenuous route and declare that I am content to provoke interest in the subject of thought experiments, and to hope that readers are more than usually indulgent.

I mentioned that there is very little literature on the subject of thought experiments. This lamentable state of affairs is about to change radically. Two other books are soon to be published: Horowitz and Massey (eds.), *Thought Experiments in Science and Philosophy*, and Roy Sorensen, *Thought Experiments*. I have seen some of the contributions to the first and they have played a role in my developing views. This is especially true of John Norton's excellent essay, 'Thought Experiments in Einstein's Work'. In several places below I borrow from it or argue against it. Norton's is one of the most intelligent and persuasive pieces going on this subject. Unfortunately, I found out about Roy Sorensen's book too late to let it have any impact on this one—though it certainly would have if I'd read it earlier. It's a rich, readable, and wide-ranging work, bound to be influential over the long haul. If an antidote to my gung-ho Platonism should be needed, then it can be found in either Norton's empiricism or Sorensen's naturalism. Both are warmly recommended.

Much of the material in this book was presented to various audiences in Canada, Dubrovnik, Yugoslavia, Lanzhou, China, and Moscow. In every case I am grateful to my hosts and numerous critics. Some of this work stems from earlier essays: 'Thought Experiments since the Scientific Revolution', *International Studies in the Philosophy of Science*, 1986, 'Einstein's Brand of Verificationism', *International Studies in the Philosophy of Science*, 1987, and 'π in the Sky', A. Irvine (ed.), *Physicalism in Mathematics*, Kluwer, 1989. There are a large number of individuals who deserve special mention: Igor Akchurin, Brian Baigrie, John L. Bell, Lars Bergström, Harvey Brown, Robert Butts, John Carruthers, Paul Forster, Rolf George, David Gooding, Ian Hacking, Andrew Irvine, Dominick Jenkins, Randell Keen, André Kukla, Igal Kvart, Lin Li, Ma Jin-Song, Penelope Maddy, Elena Mamchur, James McAllister, Cheryl Misak, Margaret Morrison, William Newton-Smith, John Norton, Kathleen Okruhlik, David Papineau, Kent Peacock, Michael Ruse, David Savan, Valerie Schweitzer, William Seager, Roy Sorensen, Demetra Sfendoni-Mentzou, Jacek Urbaniec, Alasdair Urquhart, Wang Jian-Hua, Kathleen Wilkes, and Polly Winsor. I'm grateful to them all. Finally, I am extremely grateful to David Kotchan who did the diagrams.

Note to paperback edition
I have taken this opportunity to correct a few slips and misprints. I am especially grateful to Kent Peacock who pointed many of these out to me.

Preface (to the Second Edition)

What's new in this edition? I have made numerous changes to the existing text, mostly small, but a few have been significant. Were I starting again, I would likely do things differently, but I have resisted the temptation to do a significant rewrite. I have, however, added two new chapters (Three, on what I call counter thought experiments, and Six, on the development of inertial motion). Besides extensive additions in a handful of places, notably the last several pages of what is now chapter four and several sections in what is now chapter five. The spirit of the book is unchanged.

I dedicated this book to my wife, Kathleen Okruhlik, saying that she didn't believe a word of it. She still doesn't. Nor do many others it seems, and yet, I'm very pleased with the book's reception. After expressing their inevitable scepticism about all my major claims, reviewers were overly generous. For this I am forever grateful. I also thank those who pointed out various mistakes, big and small. I doubt they have all been fixed and I dare say new ones have been introduced.

I also owe thanks to a great many people for all that I have learned from them about thought experiments. The views of some will be found in the pages to follow, but because this is a second edition, their influence does not show up in the way it would were I starting over. I will mention John Norton, however, since he constantly comes to mind when I think about these issues. Our views couldn't be farther apart, yet he pops up as the imagined audience whenever I try to formulate a good argument (or disguise a bad one). I vividly recall his words when we met for the first time shortly after we both started working on this topic: "I'm glad you exist; now I have someone to quote. Before, people accused me of attacking a strawman." Thus are some friendships born.

The new chapter six was co-authored with Mark Shumelda, who kindly gave me permission to include it here. Eran Tal did the Index to this edition. I am grateful to both of them as well as to the Social Science and Humanities Research Council of Canada for its very generous and long-standing support. SSHRC's financial aid for the philosophy community in Canada has been of the utmost value. Finally, much of this second edition was prepared while enjoying a month's stay at one of the world's academic treasures, the Rockefeller Institute in Bellagio, on Lake Como, Italy. I'm very grateful to them for allowing me to live and socialize with a very stimulating crowd and to gaze at one of the most beautiful sights imaginable, while actually getting some work done.

1 Illustrations from the laboratory of the mind

Thought experiments are performed in the laboratory of the mind. Beyond that bit of metaphor it's hard to say just what they are. We recognize them when we see them: they are visualizable; they involve mental manipulations; they are not the mere consequence of a theory-based calculation; they are often (but not always) impossible to implement as real experiments either because we lack the relevant technology or because they are simply impossible in principle, as when frictionless planes or a universe devoid of all matter are involved. If we are ever lucky enough to come up with a sharp definition of thought experiment, it is likely to be at the end of a long investigation. For now it is best to delimit our subject matter by simply giving examples; hence this chapter called 'Illustrations'. And since the examples are so exquisitely wonderful, we should want to savour them anyway, whether we have a sharp definition of thought experiment or not.

Galileo on falling bodies

Let's start with the best (i.e., my favourite). This is Galileo's wonderful argument in the *Discorsi* to show that all bodies, regardless of their weight, fall at the same speed (Galileo 1974, 66f.). It begins by noting Aristotle's view that heavier bodies fall faster than light ones (H > L). We are then asked to imagine that a heavy cannon ball is attached to a light musket ball. What would happen if they were released together?

Reasoning in the Aristotelian manner leads to an absurd conclusion. First, the light ball will slow up the heavy one (acting as a kind of drag), so the speed of the combined system would be slower than the speed of the heavy ball falling alone (H > H + L). On the other hand, the combined system is heavier than the heavy ball alone, so it should fall faster (H + L > H). We now have the absurd consequence that the heavy

Figure 1

ball is both faster and slower than the even heavier combined system. Thus, the Aristotelian theory of falling bodies is destroyed.

But the question remains, 'Which falls fastest?' The right answer is now plain as day. The paradox is resolved by making them equal; they all fall at the same speed $(H = L = H + L)$.

With the exception of Einstein, Galileo has no equal as a thought experimenter. The historian Alexandre Koyré once remarked 'Good physics is made *a priori*' (1968, 88), and he claimed for Galileo 'the glory and the merit of having known how to dispense with [real] experiments' (1968, 75). An exaggeration, no doubt, but hard to resist. Galileo, himself, couldn't resist it. On a different occasion in the *Dialogo*, Simplicio, the mouthpiece for Aristotelian physics, curtly asks Salviati, Galileo's stand-in, 'So you have not made a hundred tests, or even one? And yet you so freely declare it to be certain?' Salviati replies, 'Without experiment, I am sure that the effect will happen as I tell you, because it must happen that way' (Galileo 1967, 145).

What wonderful arrogance—and, as we shall see, so utterly justified.

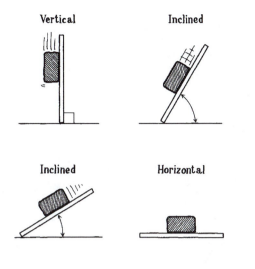

Figure 2

Stevin on the inclined plane

Suppose we have a weight resting on a plane. It is easy to tell what will happen if the plane is vertical (the weight will freely fall) or if the plane is horizontal (the weight will remain at rest). But what will happen in the intermediate cases?

Simon Stevin (1548–1620) established a number of properties of the inclined plane; one of his greatest achievements was the result of an ingenious bit of reasoning. Consider a prism-like pair of inclined (frictionless) planes with linked weights such as a chain draped over it. How will the chain move? (See Figures 3 and 4.)[1]

There are three possibilities: It will remain at rest; it will move to the left, perhaps because there is more mass on that side; it will move to the right, perhaps because the slope is steeper on that side. Stevin's answer is the first: it will remain in static equilibrium. The second diagram below clearly indicates why. By adding the links at the bottom we make a closed loop which would rotate if the force on the left were not balanced by the force on the right. Thus, we would have made a perpetual motion machine, which is presumably impossible. (The grand conclusion for mechanics drawn from this thought experiment is that when we have inclined planes of equal height then equal weights will act inversely proportional to the lengths of the planes).

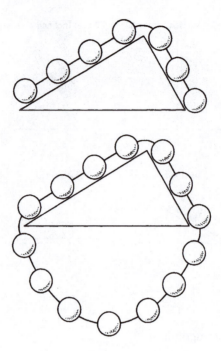

Figure 3

The assumption of no perpetual motion machines is central to the argument, not only from a logical point of view, but perhaps psychologically as well. Ernst Mach, whose beautiful account of Stevin I have followed, remarks:

> Unquestionably in the assumption from which Stevin starts, that the endless chain does not move, there is contained primarily only a *purely instinctive* cognition. He feels at once, and we with him, that we have never observed anything like a motion of the kind referred to, that a thing of such a character does not exist . . . . That Stevin ascribes to instinctive knowledge of this sort a higher authority than to simple, manifest, direct observation might excite in us astonishment if we did not ourselves possess the same inclination . . . . [But,] the instinctive is just as fallible as the distinctly conscious. Its only value is in provinces with which we are very familiar.
>
> (1960, 34ff.)

Richard Feynman, in his *Lectures on Physics,* derives some results concerning static equilibrium using the principle of virtual work. He

Figure 4

remarks, 'Cleverness, however, is relative. It can be done in a way which is even more brilliant, discovered by Stevin and inscribed on his tombstone . . . . If you get an epitaph like that on your gravestone, you are doing fine' (Feynman 1963, vol. I, ch. 4, 4f.).

Newton on centripetal force and planetary motion

Part of the Newtonian synthesis was to link the fall of an apple with the motion of the moon. How these are really the same thing is brought out in the beautiful example discussed and illustrated with a diagram late in the *Principia*.

   Newton begins by pointing out a few commonplaces we are happy to accept.

> [A] stone that is projected is by the pressure of its own weight forced out of the rectilinear path, which by the initial projection alone it should have pursued, and made to describe a curved line in the air; and through that crooked way is at last brought down to the

ground; and the greater the velocity is with which it is projected, the farther it goes before it falls to the earth. We may therefore suppose the velocity to be so increased, that it would describe an arc of 1, 2, 5, 10, 100, 1000 miles before it arrived at the earth, till at last, exceeding the limits of the earth, it should pass into space without touching it.

Newton then draws the moral for planets.

But if we now imagine bodies to be projected in the directions of lines parallel to the horizon from greater heights, as of 5, 10, 100, 1000, or more miles, or rather as many semidiameters of the earth, those bodies, according to their different velocity, and the different force of gravity in different heights, will describe arcs either concentric with the earth, or variously eccentric, and go on revolving through the heavens in those orbits just as the planets do in their orbits.

(*Principia*, 551f.)

As thought experiments go, this one probably doesn't do any work from a physics point of view; that is, Newton already had derived the motion

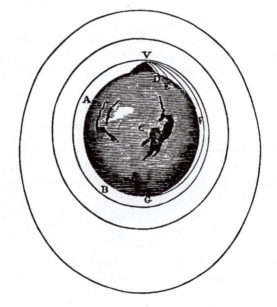

Figure 5

of a body under a central force, a derivation which applied equally to apples and the moon. What the thought experiment does do, however, is give us that 'aha' feeling, that wonderful sense of understanding what is really going on. With such an intuitive understanding of the physics involved we can often tell what is going to happen in a new situation without making explicit calculations. The physicist John Wheeler once propounded 'Wheeler's first moral principle: *Never make a calculation until you know the answer*' (Taylor and Wheeler 1963, 60). Newton's thought experiment makes this possible.

Newton's bucket and absolute space

Newton's bucket experiment which is intended to show the existence of absolute space is one of the most celebrated and notorious examples in the history of thought experiments.

Absolute space is characterized by Newton as follows: 'Absolute space, in its own nature, without relation to anything external, remains always similar and immovable. Relative space is some movable dimension or measure of the absolute spaces; which our senses determine by its position to bodies . . .' (*Principia*, 6). Given this, the characterization of motion is straightforward. 'Absolute motion is the translation of a body from one absolute place into another; and relative motion, the translation from one relative place into another' (*Principia*, 7).

Newton's view should be contrasted with, say, Leibniz's, where space is a relation among bodies. If there were no material bodies then there would be no space, according to a Leibnizian relationalist; but for Newton there is nothing incoherent about the idea of empty space. Of course, we cannot perceive absolute space; we can only perceive relative space, by perceiving the relative positions of bodies. But we must not confuse the two; those who do 'defile the purity of mathematical and philosophical truths', says Newton, when they 'confound real quantities with their relations and sensible measures' (*Principia*, 11).

Newton's actual discussion of the bucket is not as straightforward as his discussion of the globes which immediately follows in the *Principia*. So I'll first quote him on the globes example, then give a reconstructed version of the bucket.

> It is indeed a matter of great difficulty to discover . . . the true motions of particular bodies from the apparent; because the parts of that immovable space . . . by no means come under the observation of our senses. Yet the thing is not altogether desperate. For instance, if two globes, kept at a distance one from the other

by means of a cord that connects them, were revolved around their common centre of gravity, we might, from the tension of the cord, discover the endeavour of the globes to recede from the axis of their motion . . . . And thus we might find both the quantity and the determination of this circular motion, even in an immense vacuum, where there was nothing external or sensible with which the globes could be compared. But now, if in that space some remote bodies were placed that kept always position one to another, as the fixed stars do in our regions, we could not indeed determine from the relative translation of the globes among those bodies, whether the motion did belong to the globes or to the bodies. But if we observed the cord, and found that its tension was that very tension which the motions of the globes required, we might conclude the motion to be in the globes, and the bodies to be at rest.

(*Principia,* 12)

Now to the (slightly reconstructed) bucket experiment. Imagine the rest of the physical universe gone, only a solitary bucket partly filled with water remaining. The bucket is suspended by a twisted rope—don't ask what it's tied to—and released. As the rope unwinds we notice distinct states of the water/bucket system.

In state I, at the instant the bucket is released, there is no relative motion between the water and the bucket; moreover, the surface of the water is level. In state II, shortly after the bucket is released, the water and the bucket are in relative motion, that is, in motion with respect to one another. The water is still level in state II. We reach state III after some time has passed; the water and bucket are at relative rest, that is, at rest with respect to one another. But in this third state the water is not level; its surface is now concave.

The problem now is this: How do we account for the difference between state I and state III? We cannot explain it by appealing to the relative motion of the water to the bucket, since there is no relative motion in either case. Newton's answer—his explanation—is simple and profound: In state I the water and the bucket are at absolute rest (i.e., at rest with respect to absolute space) and in state III the water and bucket are in absolute motion (i.e., in motion with respect to absolute space). It is this difference in absolute motion which explains the observed difference in the shape of the water surface. On the assumption that it offers the best explanation, we should now accept the existence of absolute space.

Figure 6

Time without change?

The counterpart to absolute space is absolute time. Relationalists such as Leibniz would say that time is nothing but the sequence of events and if there were no events there would be no time. Time without change is impossible. Newton, the absolutist, had no problem with the idea of time without events. An infinite amount of time passed even though nothing happened; then God decided to create the universe. Leibniz mocked this, claiming (as he did in the case of space) that God, being perfectly rational, would not be able to choose one temporal point for creation rather than any other, since they are all the same.

More recent objections to absolute space have turned more on evidential considerations—how could we tell that time could pass without anything happening? It would seem at a minimum that something like a clock would be needed. It would record the passage of, say, one hour while no events occurred. But the ticking clock is itself a sequence of events, so such a measurement is impossible. Absolute time is in danger of being dismissed as an empty notion that has no consequences for experience whatsoever.

Sidney Shoemaker (1969) addressed this problem, not by trying to show that there is absolute time, but by trying to show that there could be a situation where we would have reasonable evidence of absolute time.

Imagine a universe with three regions (separate planets, if you like) that can see one other. Every three years planet A freezes, that is everything stops: falling objects stop in mid-flight, people stop talking mid-sentence, and so on. People on planets B and C observe this, but otherwise things continue normally in those two regions. Because there is a continuity of events in B and C, it would be perfectly correct to say that time passes on planet A, even though that region is frozen. Moreover, the people on planet A would agree that they were frozen for one year. For one thing, people on B and C could tell them and, perhaps more importantly, the people on A would notice the radical discontinuity of their experience of B and C. That is, things elsewhere would have changed considerably from before to after the freeze.

Planet A is not the only one to freeze. B freezes every four years. People on A and C observe this, pass on the information to people on B who also notice a radical discontinuity in their experience of A and C. Again, as in the first case, since events are happening elsewhere in the universe, there is no problem about the passage of time everywhere during the freeze.

Finally, planet C freezes every five years. As with the other two cases, the people on C are convinced of this. So, now we have a situation in which people on each of the planets believe that their planet freezes and that the other planets freeze and that time passes during these freezing. There is no challenge to the thesis that time requires change, since during a freeze on any one planet, there are events continuously happening elsewhere.

So far, so good. But something odd happens every 60 years. No freeze is observed on any planet. Why? There are two possible explanations that should occur to everyone:

Hypothesis 1: All the planets freeze simultaneously, since $60 = 3 \times 4 \times 5$, and since everyone is frozen, no one can observe it.

Hypothesis 2: Planet A freezes every three years, but skips a freeze every 20 cycles; B freezes every four years, but skips a freeze every 15 cycles; and C freezes every five years but skips a freeze every 12 cycles. Thus, there is no freeze at all every 60 years, which is why none is observed.

Which of these two explanations is the better one? The first is clearly better; the second is ad hoc and smacks of artificiality. The conclusion Shoemaker wants us to draw is that there are situations where the empirical evidence would strongly favour the existence of absolute time, that is, that time passes even though nothing happens.

It is not an argument that there is evidence for absolute time. It only tries to make the case that there *could be evidence* for absolute time. One might wonder, Why bother?, since such a universe of three planets as Shoemaker describes has nothing to do with us? It's a modal argument, something of a specialty among philosophers. These are cases where the notions of possibility and necessity are central, even if they are only implicit. Consider the different senses of necessity in the following three sentences:

1. Necessarily, if I have two apples and you have two apples, then together we have four apples.
2. Necessarily, if a body is unsupported, then it will fall at a rate of 9.8 m/s^2.
3. Necessarily, there can be nothing in experience that would support the existence of time without change.

The first is a mathematical necessity, which seems to hold always and everywhere. The second necessity is physical and rather limited at that. We think the laws could be different—masses might repel rather than attract—and even when the laws hold, other contingencies matter, e.g., well above the surface of the earth or on the moon the rate of fall would be much less.

The status of necessity in the third example is harder to figure out. Those who make that sort of claim (e.g., Leibniz) would consider it a kind of conceptual necessity, which like mathematical necessity would hold always and everywhere. Shoemaker, if his thought experiment is successful, has undermined this rather strong claim. He has given us an instance where it does not hold. That means it is not a necessity after all. Of course, it does not settle the issue of absolute time for us, but it does do something very important. It opens the possibility that some sort of observational evidence just might be available to us. We are no longer bound by the belief that looking for such a thing is hopeless, a conceptual impossibility. And we are no longer bowled over by the assertion: "The passage of time without any change is in principle undetectable by experience, so the very idea should be tossed out." Now it seems more like how we would respond to "You can't see electrons or phlogiston, so they don't exist." True, we'd say, but there might be other considerations that tip the balance in favour (or against) them.

Euclidean geometry

Before the rise of non-Euclidean geometry in the last century, it was a commonplace to think that geometrical reasoning yielded results about the real physical world, results known *a priori*.

> Geometry, throughout the 17th and 18th centuries, remained, in the war against empiricism, an impregnable fortress of the idealists. Those who held—as was generally held on the Continent—that certain knowledge, independent of experience, was possible about the real world, had only to point to Geometry: none but a madman, they said, would throw doubt on its validity, and none but a fool would deny its objective reference.
>
> (Russell 1897, 1)

Consequently, we can see the results of Euclidean geometry (at least those produced before the rise of non-Euclidean geometry) as a vast collection of thought experiments. This is especially so if geometry is thought of in a constructive way.[2] For example, the first of Euclid's postulates can be expressed either as 'Between any two points there exists a line' or as 'Between any two points a line can be constructed'. Euclid himself used constructive language in expressing his postulates and results, but it is unknown how philosophically seriously he intended his constructivist language to be taken. Kant, however, was adamant— geometric objects are literally human constructions, they do not have an independent existence. A theorem of Euclidean geometry is then a kind of report of an actual construction carried out in the imagination. Of course, we now consider Euclidean geometry to be false, but that doesn't detract from the fact that its postulates and theorems were a non-empirical attempt to describe the physical world, and when understood constructively, plausible examples of thought experiments.

Euclidean geometry, as an attempted *a priori* description of reality, may have faded into history, but the grand tradition of thought experiments in geometry lives on, as the next example illustrates.

Poincaré and Reichenbach on geometry

Is physical space Euclidean or not, and how could we tell? It is often claimed that our spatial experience is necessarily Euclidean. But, if this is so it is only by confining experience to the instantaneous snap-shot. If we extend our considerations to diachronic experience, a series of snap-shots, then it may be that fitting all these instantaneous experiences together can only be done in a non-Euclidean framework. Thus it might be that experience would justify the choice of a non-Euclidean geometry after all.

Henri Poincaré and Hans Reichenbach, however, tried to undermine all of this with a thought experiment about beings on a plane (i.e., a two-dimensional surface) and the choices of geometry made by these flatlanders to describe their universe.

> Let us imagine to ourselves a world only peopled with beings of no thickness, and suppose these 'infinitely flat' animals are all in one and the same plane, from which they cannot emerge. Let us further admit that this world is sufficiently distant from other worlds to be withdrawn from their influence, and while we are making these hypotheses it will not cost us much to endow these beings with reasoning power, and to believe them capable of making a geometry. In that case they will certainly attribute to space only two dimensions. But now suppose that these imaginary animals, while remaining without thickness, have the form of a spherical, and not of a plane figure, and are all on the same sphere,

Figure 7

from which they cannot escape. What kind of geometry will they construct? In the first place, it is clear that they will attribute to space only two dimensions. The straight line to them will be the shortest distance from one point on the sphere to another—that is to say, an arc of a great circle. In a word, their geometry will be spherical geometry.

(Poincaré 1952b, 37f.)

So far, so good; the two-dimensional beings aren't presenting us with any surprises. But Poincaré notes the link between metric and geometry; that is, our definition of distance (and many different ones are possible) determines the nature of the geometry.

Reichenbach's version of the thought experiment is especially clear (Reichenbach 1957, 10–12). We imagine two sets of our little beings each confined to one of the two surfaces, G or E; where G is a hemisphere and E is a plane (See Figure 8). Measuring in ordinary ways, the beings on top discover their universe has a non-Euclidean geometry. The beings at the bottom would discover their universe to be Euclidean if they, too, measured in the ordinary way. However, we shall suppose that they adopt a definition of congruence (i.e., same distance) that is equivalent to the following: Two intervals are the same when they are projections from congruent intervals from G.

Moving from right to left, they would 'discover' that their measuring rods are expanding. (They would be likely to postulate some sort of force, like heat, that has this effect on measuring rods.) Most importantly, they would 'discover' that their geometry is also non-Euclidean.

The thought experiment just sets the stage for the ensuing argument. In reality, there is no God's-eye point of view from which we can say 'The real geometry is . . .'. Poincaré (implicitly) and Reichenbach (explicitly) add the doctrine of verificationism, a philosophical view which links truth to evidence: A proposition is true (false) in so far as it can be empirically verified (refuted), but it has no truth value if it cannot be empirically tested. Since the geometry of space depends on the choice of definition of length, and that definition is conventional, it follows that geometry is conventional.

The geometrical axioms are therefore neither synthetic *a priori* intuitions nor experimental facts. They are conventions . . . . In other words, the axioms of geometry . . . are only definitions in disguise. What then are we to think of the question: Is Euclidean geometry true? It has no meaning.

(Poincaré 1952b, 50)

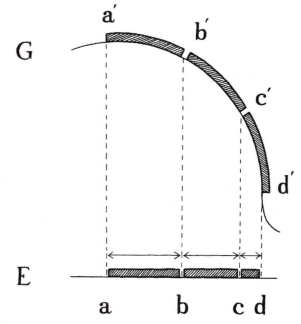

Figure 8

The jury is still out on all of this. Adolf Grünbaum, the dean of space-time philosophers, rejects the verificationism of Poincaré and Reichenbach in his *The Philosophy of Space and Time* (1974), but accepts the conventionalist conclusion. Michael Friedman, on the other hand, takes up the cudgels for a realist, nonconventional account of geometry in his impressive *Foundations of Space-Time Theories* (1983). The debate is on-going.

Non-examples

There are things that are called thought experiments, but aren't, at least not in my sense. Often in psychology or linguistics people are asked what they think about such and such. For example, someone might be asked to consider the sentence

Colourless green ideas sleep furiously.

and to decide whether it is grammatically in order. Such a process, naturally enough, is often called a thought experiment. However, it is not a thought experiment as I am considering them here; rather it is a *real* experiment about thinking. The object of the psycho-linguistic experiment is thought itself, whereas the object of a thought experiment (in my sense) is the external world and thinking is the *method* of learning something about it.

Of course, psycho-linguistic thought experiments are legitimate and important—the issue is merely terminological. No one has a monopoly on the expression 'thought experiment', but let's be sure to keep these quite distinct uses apart.

Einstein chases a light beam

According to Maxwell's theory of electrodynamics, light is an oscillation in the electromagnetic field. Maxwell's theory says that a *changing* electric field gives rise to a magnetic field, and a *changing* magnetic field gives rise to an electric field. If a charge is jiggled, it changes the electric field which creates a magnetic field which in turn creates an electric field, and so on. Maxwell's great discovery was that the wave travelling through the electromagnetic field with velocity c is light.

When he was only sixteen Einstein wondered what it would be like to run so fast as to be able to catch up to the front of a beam of light. Perhaps it would be like running toward the shore from the end of a pier stretched out into the ocean with a wave coming in: there would be a hump in the water that is stationary with respect to the runner. However, it can't be like that since change is essential for a light wave; if either the electric or the magnetic field is static it will not give rise to the other and hence there will be no electromagnetic wave.

> If I pursue a beam of light with the velocity c (velocity of light in a vacuum), I should observe such a beam of light as a spatially oscillatory electromagnetic field at rest. However, there seems to be no such thing, whether on the basis of experience or according to Maxwell's equations.
>
> (Einstein 1949, 53)

Conceptual considerations such as those brought on by this bit of youthful cleverness played a much greater role in the genesis of special relativity than worries about the Michelson-Morley experiment. Einstein goes on to describe the role of his thought experiment in later developments.

> From the very beginning it appeared to me intuitively clear that, judged from the standpoint of such an observer, everything would

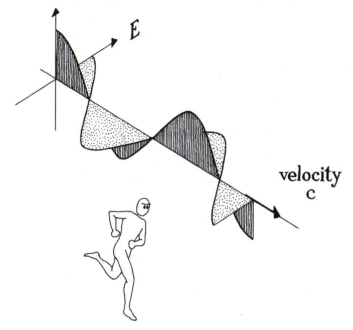

Figure 9

have to happen according to the same laws as for an observer who, relative to the earth, was at rest. For how, otherwise, should the first observer know, i.e., be able to determine, that he is in a state of fast uniform motion.

One sees that in this paradox the germ of the special relativity theory is already contained.

(1949, 53)

Seeing and manipulating

Sight is perhaps our most important sense and we have undoubtedly let this condition our thought experiments, as well. I have made being 'visualizable' or 'picturable' a hallmark of any thought experiment. Perhaps 'sensory' would be a more accurate term. After all, there is no reason why a thought experiment couldn't be about imagined sounds,[3] tastes, or smells. What is important is that it be experiencable in some way or other.

As well as being sensory, thought experiments are like real experiments in that something often gets manipulated: the balls are joined together, the links are extended and joined under the inclined plane, the observer runs to catch up to the front of the light beam. As thought experimenters, we are not so much passive observers as we are active interveners in our own imaginings. We are doubly active; active in the sense of imaginative (but this is obvious), and active in the sense of imagining ourselves to be actively manipulating (rather than passively observing) our imaginary situation.[4]

Lots of important things take place in thought which have nothing to do with thought experiments. For example, quantum electrodynamics is considered by some to be logically inconsistent. The proof of this involves showing that a particular infinite series is divergent. Such a proof is not a thought experiment–it involves neither a picturable state of affairs nor any sort of manipulation—though it is potentially as destructive as any of the examples which are.

Einstein's elevator

Special relativity brilliantly resolved some of the tensions that existed between classical mechanics and electrodynamics. However, it accounted for inertial motion only; hence the search was on for the General Theory of Relativity, a theory that included accelerated motions as well. That such a theory must link gravity to acceleration is brought out by the elevator thought experiment which is right at the heart of general relativity.

Following Einstein and Infeld, imagine that there is an inertial CS (co-ordinate system) and an elevator which is being pulled upward in CS with a constant force.

> Since the laws of mechanics are valid in this CS, the whole elevator moves with a constant acceleration in the direction of the motion. Again we listen to the explanation of the phenomena going on in the elevator and given by both the outside and inside observers.
>
> *The outside observer:* My CS is an inertial one. The elevator moves with constant acceleration, because a constant force is acting. The observers inside are in absolute motion, for them the laws of mechanics are invalid. They do not find that bodies, on which no forces are acting, are at rest. If a body is left free, it soon collides with the floor of the elevator, since the floor moves upward toward the body . . . .
>
> *The inside observer:* I do not see any reason for believing that my elevator is in absolute motion. I agree that my CS, rigidly connected with my elevator, is not really inertial, but I do not believe that

it has anything to do with absolute motion. My watch, my handkerchief, and all bodies are falling because the whole elevator is in a gravitational field.

(Einstein and Infeld 1938, 218f.)

Our thought experimenters, Einstein and Infeld, then raise the possibility of determining which of these two accounts is the right one. We are to imagine that a light ray enters the elevator horizontally through a side window and reaches the opposite wall.

The outside observer, believing in accelerated motion of the elevator, would argue: The light ray enters the window and moves horizontally, along a straight line and with constant velocity, toward the opposite wall. But the elevator moves upward and during the time in which the light travels toward the wall, the elevator changes its position. Therefore, the ray will meet a point not exactly opposite its point of entrance, but a little below . . . the light ray travels, relative to the elevator, not along a straight, but along a slightly curved line.

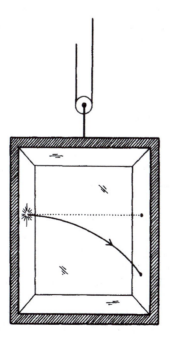

Figure 10

The inside observer, who believes in the gravitational field acting on all objects in his elevator, would say: there is no accelerated motion of the observer, but only the action of the gravitational field. A beam of light is weightless and, therefore, will not be affected by the gravitational field. If sent in a horizontal direction, it will meet the wall at a point exactly opposite to that at which it entered.

(1938, 220)

But Einstein and Infeld immediately destroy this line of reasoning.

A beam of light carries energy and energy has mass. But every inertial mass is attracted by the gravitational field as inertial and gravitational masses are equivalent. A beam of light will bend in a gravitational field exactly as a body would if thrown horizontally with a velocity equal to that of light.

(1938, 221)

Einstein and Infeld finally draw the moral from this thought experiment.

The ghosts of absolute motion and inertial CS can be expelled from physics and a new relativistic physics built. Our idealized experiments show how the problem of the general relativity theory is closely connected with that of gravitation and why the equivalence of gravitational and inertial mass is so essential for this connection.

(1938, 222)

Heisenberg's γ-ray microscope

In the early days of quantum mechanics Werner Heisenberg came up with an important inequality which now bears his name, the Heisenberg uncertainty principle. It says that the product of the uncertainties (or indeterminacies) in position and momentum is equal or greater than Planck's (reduced) constant, i.e., $\Delta p \Delta q \geq h/2\pi$. What the relation means has long been a hotly debated topic, but there is no doubt that the relation can be formally derived in a straightforward way from the first principles of the quantum theory. Heisenberg, himself, gleaned the relation from a famous and highly influential thought experiment.

Let's begin with Heisenberg's gloss on the principle itself.

The uncertainty principle refers to the degree of indeterminateness in the possible present knowledge of the simultaneous values of various quantities with which the quantum theory deals; it does not restrict, for example, the exactness of a position measurement alone. Thus suppose that the velocity of a free electron is precisely known, while the position is completely unknown. Then the principle states that every subsequent observation of the position will alter the momentum by an unknown and undeterminable amount such that after carrying out the experiment our knowledge of the electron motion is restricted by the uncertainty relation. This may be expressed in concise and general terms by saying that every experiment destroys some of the knowledge of the system which was obtained by previous experiments.

(1930, 20)

Why should we think there is this limitation on our knowledge? It comes from the following, the gamma-ray microscope thought experiment.

As a first example of the destruction of the knowledge of a particle's momentum by an apparatus determining its position, we consider the use of a microscope. Let the particle be moving at such a distance from the microscope that the cone of rays scattered from it through the objective has an angular opening ε. If λ, is the wave-length of the light illuminating it, then the uncertainty in the measurement of the x-co-ordinate according to the laws of optics governing the resolving power of any instrument is:

$$x = \frac{\lambda}{\sin \varepsilon}$$

But, for any measurement to be possible at least one photon must be scattered from the electron and pass through the microscope to the

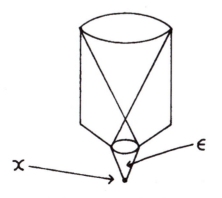

Figure 11

eye of the observer. From this photon the electron receives a Compton recoil of order of magnitude h/λ. The recoil cannot be exactly known, since the direction of the scattered photon is undetermined within the bundle of rays entering the microscope. Thus there is an uncertainty of the recoil in the x-direction of amount

$$\Delta p_x \sim (h/\lambda) \sin \varepsilon$$

and it follows that for their motion after the experiment

$$\Delta p_x \Delta x \sim h.$$

(Heisenberg 1930, 21)

One might wonder what all the fuss is about with the uncertainty principle. What has it got to do with 'observers creating reality' and other such things that are so often linked with this thought experiment? Won't it be the case that an electron will have a trajectory (i.e., a position and a momentum at all times) and we just can't know what it is? Heisenberg thinks not. The larger anti-realist morals that are sometimes drawn require the extra philosophical assumption of verificationism or operationalism which connects facts of the matter to what we can know in principle. 'If one wants to clarify what is meant by "position of an object", for example, of an electron, he has to describe an experiment by which the "position of an electron" can be measured; otherwise the term has no meaning at all' (Heisenberg, quoted in Jammer 1974, 58). Since no experiment can determine a trajectory it is a nonsense notion, according to Heisenberg, and it is meaningless to say an electron has one.

No realist would ever say 'X is unknowable, therefore X does not exist.' But the passage from epistemology (what we can know) to ontology (truth, existence) is at the heart of verificationism. As I mentioned, Heisenberg's example has been an extremely influential thought experiment. But the moral so often drawn may rest more on the philosophical assumption of verificationism or operationalism than on the details of the thought experiment itself.

Rudolf Peierls tells an interesting anecdote about the background to the thought experiment.

When Heisenberg, then a student in Munich, submitted his PhD dissertation, he was already known as a young man of outstanding ability. But he had aroused the displeasure of W. Wien, the professor of experimental physics, by not taking the laboratory classes seriously enough. It was then part of the requirements for the PhD to take a quite searching oral examination in the relevant subjects.

When Heisenberg submitted himself to questioning by Wien, the first question related to the resolving power of a microscope, and the candidate did not know the answer. The next question was about the resolving power of a telescope, and he still did not know. There were more questions about optics and no answers, and the professor decided the candidate should fail. However, a joint mark had to be returned for both experimental and theoretical physics, and after difficult negotiations between A. Sommerfeld, the theoretical physicist, and Wien, Heisenberg passed in physics with the lowest pass mark.

Four years later he wrote his famous paper about the uncertainty principle . . . [employing] a hypothetical microscope using γ-rays so as to improve the resolving power. . .

. . . the story provides an amusing illustration of the fact that even for a great man a sound knowledge of old fashioned physics can be essential.

(Peierls 1979, 34ff.)

Schrödinger's cat

Many of the great thought experiments associated with quantum mechanics are attempts to undermine the uncertainty principle, or more generally, the orthodox interpretation of the quantum formalism known as the Copenhagen interpretation. Schrödinger's cat and EPR, to be discussed next, are two of the most famous.

In the final chapter I'll explain both the formalism of quantum mechanics and the Copenhagen interpretation in some detail; for now I'll just mention a few things very briefly to make the thought experiment intelligible. A physical system is represented by a vector ψ (the state vector or wave function) in a Hilbert space. Measurement outcomes correspond to the basis or eigenvectors of the space; the state of the system, however, may be a superposition of eigenvectors.

Measurement outcomes are always eigenvalues, magnitudes which are associated with eigenstates, not with superpositions. So a very natural question is, what is the physical meaning of a superposition, and what happens when a measurement changes a superposition to an eigenstate? A common-sense realist is tempted to say that a physical system is always some way or other—superpositions merely reflect our ignorance. The Copenhagen interpretation, on the other hand, says that in a state of superposition reality itself is indeterminate—measurement, in some sense, 'creates' reality by putting the system into one of its eigenstates.

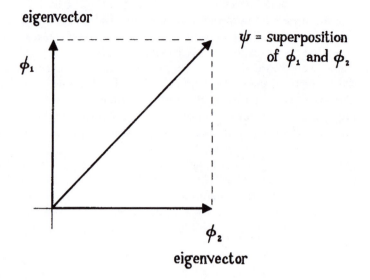

Figure 12

Schrödinger, like Einstein, detested this anti-realist view and mocked it with his famous cat paradox.

> One can even set up quite ridiculous cases. A cat is penned up in a steel chamber, along with the following diabolical device (which must be secured against direct interference by the cat): in a Geiger counter there is a tiny bit of radioactive substance, so small, that *perhaps* in the course of one hour one of the atoms decays, but also, with equal probability, perhaps none; if it happens, the counter tube discharges and through a relay releases a hammer which shatters a small flask of hydrocyanic acid. If one has left this entire system to itself for an hour, one would say that the cat still lives *if* meanwhile no atom has decayed. The first atomic decay would have poisoned it. The ψ-function of the entire system would express this by having in it the living and the dead cat (pardon the expression) mixed or smeared out in equal parts.
>
> (Schrödinger 1935, 157)

What Schrödinger has done in this example is take the indeterminacy which is strange enough in the micro-world and amplify it into the macro-world where it seems totally bizarre. On the Copenhagen interpretation, the cat would have to be in a state of superposition, a

Figure 13

mix of living and dead, until someone looked; then it would pop into one of the two possible base states.[5]

EPR

The most famous, important, and influential thought experiment directed at the anti-realist Copenhagen interpretation is the one by Einstein, Podolsky, and Rosen (1935) commonly known as EPR. (I'll present a slightly modified version due to Bohm 1951.)

EPR begins with a 'criterion of reality': if we do not interfere with a physical system in any way and we can predict an outcome of a measurement on that system with certainty, then the system must be already in the state we predicted; we are not creating the outcome. So, for example, if I can predict with certainty that there is a cup on the table in the next room, then the cup is really there independent of me. The next assumption (often called 'locality') is that measurements which are done simultaneously at a large distance from one another do not interfere with one another. (According to special relativity, two events outside each other's light cones cannot be causally connected.)

Now consider a system which decays into two photons that travel along the z-axis in opposite directions. At each end of the apparatus we shall measure the spin in the x direction, say, by holding up a polaroid filter. There are two possible outcomes for each such measurement: spin up and spin down. Moreover, the spin states are correlated: the origin has spin 0, and this is conserved; so if the left photon has spin up then the right must have spin down.

Suppose we are situated at the left detector and get the result: spin up. We can immediately infer that a measurement on the right photon will get the result: spin down. We can predict this with certainty, and since we are far removed from that measurement we are not interfering with it. Thus, the photon *already* had spin down *before* it was measured; the measurement, *contra* the Copenhagen interpretation, did *not* create the spin state; it merely discovered what was independently there.

In spite of the fact that this is an utterly beautiful and persuasive thought experiment, the majority of physicists did not accept its conclusion that the quantum theory is incomplete. Since the original EPR thought experiment J.S. Bell has taken things several steps further, so that now EPR is seen by all as fundamentally flawed (in the sense that though it is a valid argument, one of its premisses—realism or locality–must be false). But there is still no agreement on what is the right way to interpret quantum mechanics. (The final chapter takes up these matters in detail.)

Figure 14

Philosophical thought experiments

My aim is to investigate thought experiments in science, especially in physics, but since philosophy is packed full of thought experiments, it may be instructive to see some examples from this field. I'll briefly describe two recent controversial ones.

Judith Thomson (1971) has argued for the moral permissibility of abortion in spite of granting (for the sake of the argument) that the foetus is a person with a right to life. Her thought experiment consists in imagining that a great violinist has some very unusual medical condition; there is only one cure, which consists in being hooked up to *you* for nine months. Your biological make-up is the one and only one in the world which will help the violinist. In the night, unknown to you and unknown to the violinist (who is in a coma and will remain so for nine months, thus, 'innocent'), he is attached to you by a society of music lovers.

What should you do? Are you morally required to go through the nine months—an enormous sacrifice—to save the violinist's life? The answer, pretty clearly, is no. Yes, the violinist is an innocent person with a right to life and you are the one and only person in the world who can save the violinist's life; but you are not morally obliged to make the sacrifice (though you would be a moral hero if you did).

The analogy with the foetus is obvious. Thomson grants that it is an innocent person with a right to life and the pregnant mother is uniquely capable of bringing it to term. What the thought experiment does is distinguish two concepts that easily get run together: *right to life* and *right to what is needed to sustain life*. The foetus and the violinist have the former, but they do not have the latter. Having a right to life does not imply having a right to the use of another's body.[6]

Another recent and quite controversial thought experiment has to do with the nature of thought itself. AI (artificial intelligence) is the thesis that the mind is a kind of computer. The brain is the hardware, and if we could discover the software program it is running we'd then know what human understanding really is. Conversely, an ordinary computer—silicon based rather than meat based—can be truly said to think, reason, understand, and all the other cognitive states, if it is appropriately programmed. John Searle (1980) has challenged this with his 'Chinese room' thought experiment.

Imagine yourself locked in a room; you understand English, but not a word of Chinese; you are given a manual with rules in English for handling cards with Chinese writing on them. You can recognize the different Chinese words formally, that is by their shape, and when given one you must look up the appropriate rule then pass the appropriate card with some other Chinese writing on it out of the room. After some practice you would become quite adept at this, so that a native speaker of Chinese could carry on a 'conversation' with you. The Chinese room set-up could even pass the Turing test.[7]

Nevertheless, Searle thinks it is obvious nonsense to say the person in the room understands Chinese.

> it seems obvious to me in the example that I do not understand a word of the Chinese stories. I have inputs and outputs that are indistinguishable from those of the native Chinese speaker, and I can have any formal program you like, but I still understand nothing. [Any] computer for the same reasons understands nothing of any stories whether in Chinese, English, or whatever, since in the Chinese case the computer is me; and in cases where the computer is not me, the computer has nothing more than I have in the case where I understand nothing.
>
> (1980, 285)

The status of philosophical examples

What about philosophical examples? Are they really the same sort of thing as the thought examples of physics? Kathleen Wilkes (1988) doesn't think so. While allowing the legitimacy of thought experiments in science, she attacks their use in philosophy, especially in the philosophy of mind.

By way of illustration, consider a couple of examples concerning personal identity. John Locke once told a story in which the soul of a prince migrated into the body of a cobbler and vice versa. Though bizarre, the story seemed to make perfect sense. On the basis of this thought experiment Locke concluded that a person is a mental entity not a physical entity. The person in the cobbler-body is identical to the prince if and only if that person truly remembers having the prince's experiences.

Others such as Bernard Williams have cast doubt on linking personal identity to mental properties by creating thought experiments which point the other way, i.e., towards the thesis that personal identity consists in identity of body.[8] Imagine a device which is 'intended' to transplant a person A out of her own body into the body of B, and vice versa, somewhat like a computer that swaps the files on floppy disks A and B. After this process which is person A? According to Locke, A is the person who has the A-memories, which in this case will be the B-body-person. So far, so good; but now imagine the device is 'faulty' and puts the A-memories into both the A-body and the B-body. Now which is A? If we make the reasonable assumption that a person is unique (i.e., there is at most one A) then we must abandon the mental criterion for personal identity. A is the A-body-person, regardless of the memories.

Wilkes raises a number of difficulties for philosophical thought experiments: we are not given the relevant information about the background situation; the fact that we can 'imagine' something doesn't mean it's possible; what one person finds intuitively certain another will consider obviously false; but, above all, these thought experiments take us too far from the actual world—they have lost touch with reality. According to Wilkes, this is all fine for literature (when it intends to entertain) but we should not draw philosophical morals from it.

It is not that Wilkes shies away from the bizarre; much of her own theorizing is based, not on everyday examples, but rather on the behaviour of commissurotomy patients. These, however, are real folk. The people considered in philosophical thought experiments can get very weird: we are asked to imagine people splitting like amoebas, fusing like clouds, and so on. Stevin's frictionless plane, or Einstein chasing a light beam are homely by comparison. Wilkes thinks that's the real difference between them. She concedes that the dissimilarity between the legitimate thought experiment of physics and the (in her view) illegitimate thought experiment of philosophy may be a matter of degree, but notes that the difference between the bumps on her lawn and the Himalayas is also a matter of degree. She chose as a motto a delightful little poem by Hughes Mearns that sums up her view perfectly.

> As I was sitting in my chair
> I knew the bottom wasn't there,
> Nor legs nor back, but I just sat,
> Ignoring little things like that.

Fortunately, I do not have to come to grips with the challenge posed by Wilkes since I am going to deal with thought experiments in physics, all of which she considers legitimate. But I would be remiss if I didn't at least acknowledge my sense of unease.

In a pinch we might try to rescue *some* philosophical examples by invoking a distinction that Wilkes herself makes. A philosophical thought experiment is legitimate provided it does not violate the laws of nature (as they are believed to be).[9] For example, imagining people to split like an amoeba certainly violates the laws of biology. On the other hand, imagining myself as a brain-in-a-vat does not. (This thought experiment is often used to introduce worries about scepticism: how can I tell whether I'm not just a brain in a vat being subjected to computer-generated stimuli rather than having veridical experiences?) We are actually only a small technological step from being able actually to remove someone's brain, keep it alive in a vat, and give it arbitrary programmed 'experiences'. At first blush amoeba-like persons and brains in a vat may seem equally far-fetched, but on reflection they are seen to be worlds apart. Brain-in-a-vat thought experiments, on this criterion, are legitimate, since they do not violate any law of nature.

As a rule of thumb, the criterion seems useful, but it can't be exactly right. For one thing, Einstein chasing a light beam isn't nomologically possible. But there is a more important point. Too often thought experiments are used to find the laws of nature themselves; they are tools for unearthing the theoretically or nomologically possible. Stipulating the laws in advance and requiring thought experiments not to violate them would simply undermine their use as powerful tools for the investigation of nature.

Thought experiments often involve a kind of counterfactual reasoning, yet counterfactual reasoning is extremely sensitive to context. When Wilkes complains about the lack of background information in many philosophical thought experiments she is rightly noting the lack of context in which to think things through. How much background is enough? I doubt there can be a definite answer here; it's a matter of degree. I would claim, however, that there is sufficient context in each of the examples drawn from physics that I have given above. I am also inclined to think that there is enough background information to legitimize (in prinriple) the brain-in-a-vat example, as well as Thomson's violinist and Searle's Chinese room thought experiments. But, like Wilkes, I know so little about a world where people could split like an amoeba that I simply don't know what to say about it.

In spite of qualms, my attitude to various thought experiments is embodied in a remark by Lakatos:

> if we want to learn about anything really deep, we have to study it not in its 'normal', regular, usual form, but in its critical state, in fever, in passion. If you want to know the normal healthy body, study it when it is abnormal, when it is ill. If you want to know functions, study their singularities. If you want to know ordinary polyhedra, study their lunatic fringe.
>
> (1976, 23)

Other fields

Thought experiments flourish in physics and in philosophy; to some extent they are to be found in mathematics as well. But they seem to be rather scarce in the other sciences. In chemistry, for example, I can't find any at all. Biology, on the other hand, is quite rich. Darwin in *The Origin of Species* imagined giraffes with necks of different lengths. Obviously, some would fare better than others in the 'struggle for existence'. Louis Agassiz imagined a world with lobsters, but without other articulated creatures (crustaceans, spiders, and insects). He thought it was clear that even in such a situation the taxonomist would need species, genus, family, etc. to classify the solitary lobster. This in turn suggested that the taxonomer's classifications are not arbitrary or conventional. In recent times thought experiments using game theoretic principles have been widely used. I have in mind such examples as the hawk-dove games of strategy where we imagine a population consisting of these two types. Hawks always fight until either victorious or injured; doves always run away. By assigning pay-offs for various outcomes one can calculate what a stable population might look like. (See Maynard Smith, 1978, for a popular account.)[10]

Why do some fields have so many more thought experiments than others? It may have something to do with the personalities or the scientific styles of those who do physics and philosophy. Or it may be an historical accident; perhaps a tradition of doing thought experiments has grown up in some fields, but not in others.

Of course, a much more interesting explanation would link the existence of thought experiments to the nature of the subject matter. Most people hold the view that chemistry has no laws of its own; it is entirely reducible to physics. If this is so, and if laws are crucially involved in any thought experiment, then we may have the makings of an explanation for the lack of thought experiments in chemistry. But I won't pursue the issue here.

Perhaps there is something about physics and philosophy which lends itself to this kind of conceptual analysis. If we could somehow forge a link it would probably tell us a lot about physics and philosophy as well as telling us a lot about thought experiments. But I have no idea what it could be.

2 The structure of thought experiments

We can praise the inventors and savour their achievements, but can we say more about thought experiments? Are they just a curious and diverse collection of dazzling displays of mental gymnastics, each unique in its own way, or is there some pattern? The first task of any analysis of thought experiments must be the construction of a classification scheme.

Some commentators on thought experiments have suggested that they work in some one particular way or other. Not so—they work in many different ways, just as real experiments do. For example, real experiments sometimes test (i.e., confirm or refute) scientific conjectures; sometimes they illustrate theories or simulate natural phenomena; and sometimes they uncover or make new phenomena. (See Harré 1981 and Hacking 1983.) Thought experiments are at least as richly diverse in their uses as this. Nevertheless, there are some definite patterns.

The taxonomy I propose is as follows.[1] First, thought experiments break into two general kinds, which I'll call *destructive* and *constructive*, respectively. The latter kind break into three further kinds which I'll call *direct, conjectural,* and *mediative.* There is a small class of thought experiments which are simultaneously in the destructive and the constructive camps; these are the truly remarkable ones which I call *Platonic.* Their very special role in science will be described and argued for in chapter five.

Karl Popper (1959) refers to the *critical* and the *heuristic* uses of thought experiments, which corresponds roughly to my destructive and constructive types. However, given Popper's well-known views on theory confirmation—there's no such thing, according to him—we part company sharply on the positive view. John Norton (1991) classifies thought experiments as *Type I,* which are deductive arguments (whether in support of a theory or a *reductio ad absurdum* of it), and *Type II* which involve some sort of inductive inference. Like Norton,

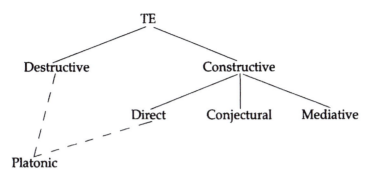

I think some thought experiments involve inductive reasoning and others deductive reasoning; but as we shall see below, after this initial agreement we also go our separate ways.

Now to some details.

Destructive thought experiments

As its name suggests, a destructive thought experiment is an argument directed against a theory. It is a picturesque *reductio ad absurdum;* it destroys or at least presents serious problems for a theory, usually by pointing out a shortcoming in its general framework. Such a problem may be anything from a minor tension with other well-entrenched theories to an outright contradiction within the theory itself. We could use this to divide destructive thought experiments into internally and externally destructive, but I will not further complicate the taxonomy here.

The example of Einstein chasing a light beam showed a problem in Maxwell's theory of light, or rather in the joint assumption of Maxwell's theory plus classical mechanics. Schrödinger's cat thought experiment did not show that quantum mechanics is logically false, but it did show that it is wildly counter-intuitive, perhaps to the point of being absurd. Galileo's falling bodies example showed that Aristotle's theory of motion was logically impossible, since it led to the contradictory conclusion that the heavy body is both faster and not faster than the coupled body.

I'll further illustrate the class of destructive thought experiments with a contrary pair. The first shows that the earth could not be in motion, *contra* the Copernican hypothesis; the second that it could. These examples have the additional merit of showing that thought experiments, even the most ingenious, are quite fallible.

The first argument put forth several times on behalf of Aristotle runs as follows. If the earth really were moving then a cannon ball dropped from a tower would not fall to the base of the tower but would fall well behind (since the tower which is attached to the spinning earth would move away); moreover, a cannon ball fired from east to west would not travel the same distance as one fired from west to east; similarly, a bird which took to flight would be left behind as the earth hurtled forward in its voyage through space.

These simple considerations were picturesque and plausible in their day. They constituted a very effective thought experiment (or set of related thought experiments) which undermined Copernicus. Galileo combated it with one of his own which established what we now refer to as Galilean relativity. (More details in chapter six.)

> Shut yourself up with some friend in the main cabin below decks on some large ship, and have with you there some flies, butterflies, and other small animals. Have a large bowl of water with some fish in it; hang up a bottle that empties drop by drop into a vessel beneath it. With the ship standing still, observe carefully how the little animals fly with equal speed to all sides of the cabin. The fish swim indifferently in all directions; the drops fall into the vessel beneath; and, in throwing something to your friend, you need throw it no more strongly in one direction than another, the distances being equal; jumping with your feet together, you pass equal spaces in every direction. When you have observed all these things carefully (though there is no doubt that when the ship is standing still everything must happen this way), have the ship proceed in any direction you like, so long as the motion is uniform and not fluctuating this way and that. You will discover not the least change in all the effects named, nor could you tell from any of them whether the ship was moving or standing still.
>
> (*Dialogo*, 186f.)

Galileo goes on to list the effects that remain the same: the jumps, the flight of the birds, the drops into the vessel below, the force required to throw something to a friend, etc.

This example looks like what I call a 'merely imagined' experiment, especially since Galileo tells us to go out and actually do it. However, it would probably be impossible to find a ship that was under sail and not 'fluctuating this way and that'; the wind that makes it move also makes waves. More importantly, we don't need to actually perform the experiment since we are sure we know (as soon as Galileo directs our attention to it) what the result must be. (Indeed in a real experiment if we experienced anything else we'd have a revolution on our hands.) So it is a genuine thought experiment and it undermines the conclusion of the earlier thought experiment which had seemed such a barrier to the acceptance of Copernicus.

Constructive thought experiments

There is room for fine tuning in the category of destructive thought experiments. We could distinguish, as I mentioned above, between those which show a theory to be internally inconsistent and those which show a theory to be in conflict with other well-entrenched beliefs. Thus, Galileo showed Aristotle's theory of free fall to be internally at fault while Schrödinger showed the quantum theory to be at odds, not with itself, but with very basic common sense.

I'll forgo such fine tuning in the case of destructive thought experiments, but not when it comes to constructive ones. Here the difference between what I'm calling direct, conjectural, and mediative is of prime importance. The structures of these three types of thought experiment are quite different even though they all aim at establishing a positive result.

Mediative thought experiments

A *mediative thought experiment* is one which facilitates a conclusion drawn from a specific, well-articulated theory. There may be many different ways in which this can be done. For example, a mediative thought experiment might illustrate some otherwise highly counter-intuitive aspect of the theory thereby making it seem more palatable; or it may act like a diagram in a geometrical proof in that it helps us to understand the formal derivation and may even have been essential in discovering the formal proof.

Maxwell's demon is a perfect example of this illustrative role. In the nineteenth century James Clerk Maxwell was urging the molecular-kinetic theory of heat (Maxwell 1871). A gas is a collection of molecules in rapid random motion and the underlying laws which govern it, said Maxwell, are Newton's. Temperature is due to the kinetic energy of the molecules; pressure is due to the molecules hitting the walls of the container, etc. Since the number of particles in any gas is enormously large, the treatment must be statistical, and here lay Maxwell's difficulty. One of the requirements for a successful statistical theory of heat is the derivation of the second law of thermodynamics which

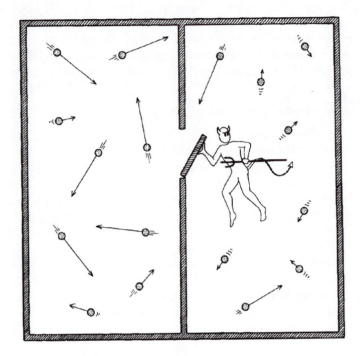

Figure 15

says: In any change of state entropy must remain the same or increase; it cannot decrease. Equivalently, heat cannot pass from a cold to a hot body. But the best any statistical law of entropy can do is make the decrease of entropy very improbable. Thus, on Maxwell's theory there is some chance (though very small) that heat would flow from a cold body to a hot body when brought into contact, something which has never been experienced and which is absolutely forbidden by classical thermodynamics.

The demon thought experiment was Maxwell's attempt to make the possible decrease of entropy in his theory not seem so obviously absurd. We are to imagine two gases (one hot and the other cold) in separate chambers brought together; there is a little door between the two containers and a little intelligent being who controls the door. Even though the average molecule in the hot gas is faster that the average in the cold, there is a distribution of molecules at various speeds in each chamber. The demon lets fast molecules from the cold gas into the hot chamber and slow molecules from the hot gas into the cold chamber.

The consequence of this is to *increase* the average speed of the molecules in the hot chamber and to *decrease* the average speed in the cold one. Of course, this just means making the hot gas hotter and the cold gas colder, violating the second law of classical thermodynamics.

The point of the whole exercise is to show that what was unthinkable is not so unthinkable after all; it is, we see on reflection, not an objection to Maxwell's version of the second law that it is statistical and allows the possibility of a decrease in entropy.

Maxwell's demon helps to make some of the conclusions of the theory more plausible; it removes a barrier to its acceptance. Other mediative thought experiments play a more instrumental role in arriving at some conclusion. The next example comes from special relativity and shows that there can be no such thing as a rigid body.

Consider a 'rigid' metre stick which is sliding over a table top which has a hole one metre in diameter; there is a gravitational force down. We reason as follows: In the table frame the moving metre stick is Lorentz contracted; hence it is shorter than one metre; thus, due to the gravitational force downward, it will fall into the hole. On the other hand, in the metre stick frame the moving table is Lorentz contracted, hence the diameter of the hole is less than one metre, hence the metre stick will pass from one side to the other without falling into the hole. This apparent paradox is resolved by noting that the metre stick would actually drop into the hole under the force of gravity. The metre stick

Figure 16

cannot be genuinely rigid, that is, maintain its shape even within its own frame of reference.

The general conclusion drawn from this thought experiment is that there are no rigid bodies in special relativity. The consequences of this are enormous and far surpass the apparent triviality of the thought experiment.

> [W]e can draw certain conclusions concerning the treatment of *'elementary'* particles, i.e., particles whose state we assume to describe completely by giving its three coordinates and the three components of its velocity as a whole. It is obvious that if an elementary particle had finite dimensions, i.e., if it were extended in space, it could not be deformable, since the concept of deformability is related to the possibility of independent motion of individual parts of the body. But . . . the theory of relativity shows that it is impossible for absolutely rigid bodies to exist.
>
> Thus we come to the conclusion that in classical (non-quantum) relativistic mechanics, we cannot ascribe finite dimensions to particles which we regard as elementary.
>
> In other words, within the framework of classical theory elementary particles must be treated as points.
>
> (Landau and Lifshitz 1975, 44)

Conjectural thought experiments

In a mediative thought experiment we start with a given background theory and the thought experiment acts like a midwife in drawing out a new conclusion. Not all thought experiments work like this; there is an important class in which we do not start from a given theory. The point of such a thought experiment is to establish some (thought-experimental) phenomenon; we then hypothesize a theory to explain that phenomenon. I shall call this kind of thought experiment *conjectural* since we are prodded into conjecturing an explanation for the events experienced in the thought experiment.

Newton's bucket is a prime example. It was described in detail earlier, so I'll only quickly review it now. We are to think away all the rest of the material universe except a bucket of water which endures three distinct, successive states:

I. There is no relative motion between bucket and water; the water surface is flat.

II. There is relative motion between water and bucket.

III. There is no relative motion between water and bucket; the water surface is concave.

This is the phenomenon; it is produced by the thought experiment and it needs to be explained. The explanation that Newton gives is this: In case III, but not in case I, the bucket and water are rotating with respect to absolute space.

I stress that absolute space is not derived from the thought experimental phenomenon; it is postulated to explain it. It is quite instructive to look at the response of Newton's critics, Berkeley and Mach. Both, of course, rejected absolute space. They dismissed the explanation offered of the phenomenon, but what is more important, they denied the phenomenon itself. That is, they denied that the water would climb the walls of the bucket in a universe without other material bodies in it. They proclaimed that the distant fixed stars are responsible for inertia and if the stars could be given a push so that they 'rotated' around the bucket, then we would see the water climb the sides as in case III. In short, according to Berkeley and Mach, it is the *relative* motion of the water/bucket system to the fixed stars that causes the water to climb the walls, not absolute space.

Einstein considered the same sort of thought experiment as Newton did, but, invoking a kind of verificationism, came to quite a different, non-absolutist conclusion. In his thought experiment he considers two globes, one a sphere and the other an ellipsoid of revolution; they are in observable rotation with respect to one another. He asked, 'What is the reason for the difference in the two bodies?' He then set philosophical conditions which rule out Newton's answer:

> No answer can be admitted as epistemically satisfactory, unless the reason given is an *observable fact of experience*. The law of causality has not the significance of a statement as to the world of experience, except when *observable facts* ultimately appear as causes and effects.
>
> (1916, 112)

This particular thought experiment as well as Einstein's views in general will be discussed in chapter seven.

Direct thought experiments

The final class of constructive thought experiments I call *direct*. They resemble mediative thought experiments in that they start with unproblematic (thought-experimental) phenomena, rather than conjectured phenomena. On the other hand, direct thought experiments, like conjectural ones, do not *start* from a given well-articulated theory—they *end* with one.

Examples of direct thought experiments which were described earlier include: Galileo on free fall, EPR, Stevin on the inclined plane, and Einstein's elevator. As another illustrative example I'll now look at one of Huygens's elegant thought experiments which established his collision laws.

In his posthumous work *De Motu Corporum ex Percussione,* Christian Huygens (1629–1695) established a number of laws governing the impact of moving bodies. His background assumptions (which were more definite than Stevin's, but not really a well-articulated theory) included the law of inertia, Descartes's relativity of motion, and the principle that elastic bodies of equal mass, colliding with equal and opposite velocities, will fly apart with equal and opposite velocities. Working with these assumptions Huygens constructed an elegant thought experiment to establish the general principle: Equal elastic masses will exchange their velocities on impact.

A boat is moving with velocity v next to an embankment. On board a collision of two equal masses takes place. The masses are moving *in the boat* with velocities v and $-v$. They rebound, of course, with velocities $-v$ and v, respectively. As well as the observer in the boat there is a second on the land. For the second the velocities are $2v$ and 0 before the collision, and 0 and $2v$ after. We may conclude from this that an elastic body, in a collision, passes on all its velocity to another of equal mass. We may also conclude the even stronger result. Since the boat may have any velocity v', the velocities (on the shore) before impact will be

Figure 17

$v' + v$ and $v' - v$ before impact, and $v' - v$ and $v' + v$ after, respectively; and since v' is arbitrary, it follows that in any impact of equal elastic masses, they will exchange their velocities.

Huygens went on to derive many more collision rules from this thought experiment, including rules for unequal masses. (For further details see Dugas 1955 or Mach 1960.)

The contrast between direct and mediative may just be a matter of degree. In a direct thought experiment one might begin with some vague general principle, like Stevin with the belief that there are no perpetual motion machines. The boundary between vague general beliefs and well-articulated theories is no doubt fuzzy, but the difference in degree in many cases is so considerable as to justify my insisting on a difference between direct and mediative thought experiments. If it is maintained that Huygens did indeed work with a well-articulated theory in deriving his results, then this thought experiment should be re-classified as mediative. But for now I'm inclined to see it as direct.

I am also inclined to claim that the difference between direct and mediative thought experiments is more than a 'mere matter of degree'. In a mediative thought experiment a logical relation (deductive or statistical, depending on the case at hand) is clarified between the theory and the conclusion. The thought experiment is a psychological help, but the logical relation exists independently of it. In a direct thought experiment, however, the logical relation between vague general principle and conclusion simply does not exist.

Though direct thought experiments do not start with a specific theory which is maintained throughout the chain of reasoning (as is, say, special relativity in the mediative thought experiment which results in showing the impossibility of rigid objects), nevertheless, some direct thought experiments start with well-articulated theories which they then go on to refute. For example, Galileo spells out Aristotle's account of free fall and then shows that it is self-contradictory. But I presume there is no ambiguity about this.

Platonic thought experiments

A small number of thought experiments fall into two categories; they are simultaneously destructive and constructive (direct). Two examples have already been given. Galileo's account of free fall did two distinct things: first, it destroyed Aristotle's view that heavier objects fall faster; and second, it established a new account that all objects fall at the same speed. Similarly, EPR did two distinct things: it destroyed the Copenhagen interpretation and it established the incompleteness of quantum mechanics. (Thought experiments are fallible, of course, so my use of terms like 'destroy' and 'establish' should be understood as merely tentative.)

Yet another example of this rare type of thought experiment is Leibniz's account of *vis viva*. When the giants were constructing what we now call classical mechanics, it was commonly agreed that something was conserved. Descartes and Leibniz agreed that *motive force* is conserved, but just what is it? Descartes took it to be *quantity of motion* (roughly, momentum as we would now call it)[2] while Leibniz took it to be *vis viva* (roughly, twice the kinetic energy). In one simple elegant example Leibniz destroyed the Cartesian view and established his own.

Leibniz (1686) starts by making assumptions that any Cartesian would agree to: first, that the quantity of force (whatever force is) acquired by a body in falling through some distance is equal to the force needed to raise it back to the height it started from, and second, the force needed to raise a one-kilogram body (A) four metres (C-D) is the same as the

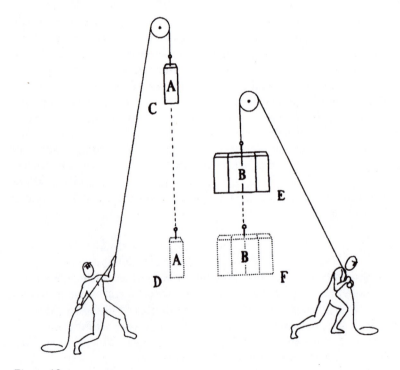

Figure 18

force needed to raise a four-kilogram body (B) one metre (E-F). From these two assumptions it follows that the force acquired by A in falling four metres is the same as the force acquired by B in one metre.

Having set the stage we can now compute the quantities of motion of each using a relation established by Galileo. The velocity of A after falling four metres will be two metres per second. Multiply this by its weight, one kilogram, and we get a quantity of motion for A of $2 \times 1 = 2$. The velocity of B after falling one metre is one metre per second, so its quantity of motion is $1 \times 4 = 4$.

This simple example refutes the Cartesian claim that force is quantity of motion. But this is merely the first step; Leibniz goes on to give us the right answer. It is elicited from the fact that the distance any body has fallen is proportional to the square of its velocity. So Leibniz's answer to the question, What is this motive force which is conserved? is *vis viva*, i.e., mv^2. It is easily verified in this or any other similar example.[3]

The Platonic class of thought experiments will obviously be the most controversial. Critics may quibble with the details of the rest of my taxonomy, but no empiricist could seriously entertain the possibility of *a priori* knowledge of nature. However, I will not argue for the existence of Platonic thought experiments now; chapter five is devoted almost entirely to this. Instead, I will first discuss other matters, one of which is a defence of Platonism in mathematics. Mathematical Platonism is an eminently respectable view, unlike a priorism about the physical world. Accepting the need for real abstract entities to explain mathematical thinking will ease the way in making Platonism in thought experiments more palatable; it will give us a model on which to base our talk of this remarkable class of thought experiments.

Theory and evidence

Having now seen a variety of examples of the three types of constructive thought experiment, we can draw one major conclusion about them:

> The burden of any constructive thought experiment consists in establishing (in the imagination) the thought-experimental phenomenon. This phenomenon then acts as fairly conclusive evidence for some theory.

In almost every thought experiment, the trick seems to be in getting the imagined phenomena going. The jump from the single case examined in the thought experiment to the theoretical conclusion is really a very small jump. Indeed, an inductive leap is often made (in the direct and conjectural cases, though not usually in the mediative), but it is certainly not the sort of thing one would be tempted to call a 'bold conjecture'.[4]

This interesting fact about thought experiments calls for some speculation. I suggest it has something to do with natural kind reasoning.[5] We hesitate to jump to the conclusion that all swans are white after seeing just one, but we don't hesitate to think that all gold has atomic number 79 after analysing only one small sample. The difference is that we are unsure whether being white is an essential property of swans (indeed, we probably doubt it) while the atomic number does seem to be an essential property of gold. Thus, if one bit of gold has this property then they all must have it. In this way natural kind properties are linked to inductive inference; the general form of inference goes: If anything of kind A has essential property B, then every A has B. In any particular case doubts can arise—but not over the conditional. What may be questioned is whether the antecedent is true: Is this really a sample of gold? Does it really have atomic number 79, or have we made a faulty chemical analysis? Is micro-structure really the essential property of any material body or is it something else such as functional role?

When we set up a thought experiment, we are really dealing with what we take to be natural kind properties. Those are the ones which are operative in the thought experiment. Now, of course, we can be mistaken in our belief that the relevant properties in the thought experiment really are natural kind properties, but as long as we are convinced that they are, then the inductive leap to some general conclusion is as easy as it is in the gold case.[6]

Norton's empiricism

John Norton has given an elegant and fairly persuasive empiricist account of thought experiments. His view does considerable justice to a wide range of thought experiments, especially those I call destructive and mediative.

Norton's basic idea is that thought experiments are just arguments; they are derivations from given premises which employ strictly irrelevant, though perhaps usefully picturesque, elements. It is of central importance in Norton's account that the premises in the argument have been established in some fashion acceptable to an empiricist. The conclusion, then, will pass empiricist muster, since the deductive or inductive consequences of empirically acceptable premises are themselves empirically acceptable. Norton's precise characterization is this:

Thought experiments are arguments which

(i) posit hypothetical or counterfactual states of affairs and
(ii) invoke particulars irrelevant to the generality of the conclusion.

It is motivated by a staunch empiricism.

> Thought experiments in physics provide or purport to provide us information about the physical world. Since they are *thought* experiments rather than *physical* experiments, this information does not come from the reporting of new empirical data. Thus there is only one non-controversial source from which this information can come: it is elicited from information we already have by an identifiable argument, although that argument might not be laid out in detail in the statement of the thought experiment. The alternative to this view is to suppose that thought experiments provide some new and even mysterious route to knowledge of the physical world.
>
> (Norton 1991, 129)

Norton's account is highly appealing, but in spite of its many virtues it fails to do justice to two types of thought experiment. It is a good account of both destructive and mediative thought experiments, but it fails when we consider either conjectural or direct thought experiments. In Norton's account we must start with clearly specified premises, a well-articulated background theory. The thought experiment is an argument; it culminates in a conclusion. We have clearly specified premises to work from in either destructive or mediative examples; but in the case of either direct or conjectural thought experiments we simply do not have a definite background theory from which we can be said to be arguing to our conclusion.

In the example of Newton's bucket, which is a conjectural thought experiment, the argument culminates in absolute space which is postulated to explain the thought-experimental phenomenon. Absolute space is not the conclusion of an argument, it is the explanation for a phenomenon that Newton, in effect, postulates. And in the case of Stevin's inclined plane which is a direct thought experiment, we establish that the chain would remain in static equilibrium. Of course, the result does not come *ex nihilo*, but we debase the idea of a theory if we say the vague ideas we have about symmetry and perpetual motion machines constitute the premises of an argument that Stevin (even in a vague sense) employed.

Reconstructing arguments is no easy matter. Saying so is not a criticism of Norton but an acknowledgement that the task he has taken on is difficult. His reconstruction of the Galileo thought experiment in the form of an argument goes as follows:

1. Assumption for *reductio* proof: the speed of fall of bodies in a given medium is proportionate to their weights.
2. From 1: if a large stone falls with 8 degrees of speed, a smaller stone half its weight will fall with 4 degrees of speed.
3. Assumption: if a slower falling stone is connected to a faster falling stone, the slower will retard the faster and the faster speed the slower.
4. From 3: if the two stones of 2 are connected, their composite will fall slower than 8 degrees of speed.
5. Assumption: the composite of two weights has greater weight than the larger.
6. From 1 and 5: the composite will fall faster than 8 degrees.
7. Conclusions 4 and 6 contradict.
8. Therefore, we must reject assumption 1.
9. Therefore, all stones fall alike.

(Norton 1996, 341–2)

The argument from [1] as far as [8] is valid. This is the destructive component of Galileo's thought experiment and Norton is quite right about being able to reconstruct it as an argument. However, the move from [8] to [9] is clearly not valid. The negation of [1] is 'It is not the case that the speed of falling bodies in a given medium is proportionate to their weight', which is certainly not the same as saying they fall at the same speed. There are many different ways that [1] could be false. For instance, as I already mentioned, they might fall in some manner connected to colour or chemical composition.

Of course, the failure of this case does not show that any other reconstruction must also fail. But the difficulty in getting this right does suggest something important that Norton ignores in his account. Even if a reconstruction of a thought experiment as an explicit argument is always possible, the difficulty in finding one and making it explicit suggests that thought experimenters are not going through a mental process that is somehow closely related to an argument. Chess masters seem to be aware of patterns and reason with them when playing. It is always possible to reconstruct what they do as merely calculating a long sequence of possible moves (as a computer actually does), but this importantly misses what exceptional chess players actually do.

When Norton says, 'The alternative to [his empiricist] view is to suppose that thought experiments provide some new and even mysterious route to knowledge of the physical world' what he has in mind and wishes to dismiss out of hand is the kind of view I outlined earlier and will argue for at length in chapter five.

3 Counter thought experiments

There are different types of thought experiments. The last chapter presented a taxonomy. Positive or constructive thought experiments support some theory, while negative or destructive ones undermine. There is much more to be said. My interest here is in a special class of negative thought experiments that I shall call *counter thought experiments*. I'll largely ignore other types in this chapter, except for the sake of contrast. The sense of counter that I intend is quite specific. It will involve the countering of the phenomenon of a thought experiment, not its set-up, nor its result. First, we need to fine-tune the category of negative thought experiment.

Negative thought experiments

As we saw in the last chapter, one type of destructive thought experiment shows some existing theory to be self-contradictory. Einstein chased a light beam with a view to seeing what the wave front looked like. If we ran beside a water wave, it would appear a static hump, so one should have a similar experience in the case of light, since light, according to Maxwell's theory, is an electromagnetic wave. The light wave, however, is dependent on change: a changing magnetic field gives rise to an electric field, and a changing electric field gives rise to a magnetic field. When Einstein catches up to the front of the light wave, he would see static fields, but then a light wave cannot exist.

Showing an internal contradiction is not the only way to undermine an existing theory. Some thought experiments show the theory to be contrary to other established (including common-sense) beliefs. Schrödinger's cat is a prime example. Schrödinger took the weirdness of the Copenhagen interpretation of quantum mechanics at the micro-level and brought it to the macro-world of everyday objects such as a cat, implying that even a cat could be in a superposition of living and dead. This is not a contradiction in the strict logical sense, but it is unquestionably a violation of common sense.

In argument terms, these thought experiments show the premisses to be false. That is, they show that the theory in question must be wrong. In the Einstein case, the thought experiments show that there is something wrong with the conjunction of electrodynamics and basic assumptions about moving reference frames. In the second the target is the Copenhagen interpretation that allows physical systems to be in reality in states of superposition.

Of course, these are debatable outcomes. As with any real experiment, there is lots of room for rival interpretations, not to mention outright mistakes. But we need not concern ourselves with this now. Our aim to understand different types of negative thought experiments.

A second type of negative thought experiment shows a situation that undermines a crucial inference. It does not challenge the premisses the way the first type of negative thought experiment does. In terms of logic, this class of thought experiment aims to show invalidity (i.e., the premisses may be true but the conclusion does not follow from them).

In *De Rerum Naturua*, Lucretius presented a thought experiment to show that space is infinite (Lucretius 1992, I, 951–68). We imagine ourselves near the alleged edge of space; we throw a spear; we see it either sail through the 'edge' or we see it bounce back. In the former case the 'edge' isn't the edge, after all. In the latter case, there must be something beyond the 'edge' that repelled the spear. Either way, the 'edge' isn't really an edge of space, after all. So, Lucretius concludes, space is infinite.

We would now make a distinction between *unbounded* and *infinite*, so that the conclusion Lucretius drew, we now clearly recognize, does not follow from what went before. The way to see the mistake is to do another thought experiment, one that starts out like the former, but comes to a very different result. The difference is that this time we imagine ourselves in a space with a different topology from the normal Euclidean. Think of the difference for two-dimensional beings between a flat surface and the surface of a sphere such as the earth. We want a three-dimensional version of this, but the thought experiment in two dimensions will suffice to make the point. The two-dimensional being on the sphere would throw their spears at any alleged edge of space and would reason just as in the initial thought experiment. They would be right to conclude that any alleged edge of space is not an edge, but they are obviously wrong to infer that space is infinite, since they are merely on the surface of a finite sphere.

Poincaré's disk people (Poincaré 1952) works in a similar way. The measuring rods of the disk people shrink as they move toward the edge of their space, so that they might even come to mistakenly think that they live in an infinite universe. If they threw spears, those spears would behave just as Lucretius says. But the space is finite.

The non-standard topology and the non-Euclidean space examples show that Lucretius' premises do not entail his conclusion. In argument terms, it is not valid. There is no blame cast on his premises, at least not on the explicit ones involving spear throwing.

The third type of negative thought experiment—the one I am chiefly concerned with here—is the *counter thought experiment*. It does not involve criticizing the premises or the inference of a thought experiment. It has a different structure. The balance of this chapter will be devoted to describing them and trying to determine some of their main properties. As examples, I will discuss three:

* Galileo against the Aristotelians (principle of relativity)
* Mach against Newton (absolute space)
* Dennett against Jackson (physicalism)

Counter thought experiments—as I have already mentioned, but wish to stress—are not readily understood in terms of the usual argument form; they are not about validity or soundness or anything analogous. They are directed at a given thought experiment, but they challenge neither the premises nor the concluding inference. Instead, counter thought experiments deny the phenomena of the initial thought experiments.

Boundaries of an experiment

Experimenters do a great many things. They set up their equipment; they let it run and see what happens; they measure; they calculate; they interpret; and they draw some conclusions which they publish. It is not easy to draw the boundaries of an experiment, that is, to say where the experiments ends and the calculating or the interpreting or the drawing conclusions from it begins. The distinction between theory and observation, for instance, is fuzzy at best and the case has been well-made that observations everywhere are theory-laden. I readily grant this, but wish to focus on something a bit more mundane, a distinction between experiment in the broad sense and in the narrow sense, as I will characterize these terms.

In the narrow sense, an experiment includes the set-up and the observation (which may be highly theory-laden). In the broad sense, the experiment includes background assumptions and initial theorizing, the set-up, observation, additional theorizing, calculating, and drawing the final conclusion. It is this final result, with an account of how it was obtained, that we read in a journal. I doubt there could be a sharp distinction drawn between narrow and broad experiment. And what goes for real experiments goes for thought experiments, too. But there is still a rough and ready distinction. The narrow part is the phenomenon—what we see. We could put this in a simple schematic way:

Theory & Background → Phenomenon → Result

The narrow sense of experiment—whether real experiment or thought experiment—is what we observe. It is the phenomenon, the middle of the schema. The broad sense includes the whole thing from theory and background assumptions to the final result.

Looking at the simple schema, it is obvious that challenges could come at different points. (Note: The arrows just mean 'is followed by', but for some purposes they might be taken loosely as deductive or inductive implications.) One could challenge the assumptions that played a role in the set-up, that is, the theoretical and other background assumptions that went into it. This is what one class of negative thought experiments do; they attack the premises. One could also challenge the inference to the final result from what went before. This is what another class of negative thought experiments do: they attack the alleged validity. But obviously, there is also a third way: one could challenge the phenomenon of the thought experiment; that is, one could claim that the phenomenon does not occur, or that what is observed is quite different from what was initially claimed. Let's illustrate this with some examples of each of these types of challenges. In the first set of examples to follow, the phenomena is never at issue; they are *not* examples of Counter Thought Experiments.

Lucretius, Searle, Thomson

Two thought experiments undermine Lucretius. They both work the same way: they accept the explicit background beliefs of Lucretius, they accept the way the thought experiment is set up, and they accept the observations, as described. They reject the conclusion. Both provide a situation where the background and the phenomenon are as Lucretius wants, but the conclusion of infinite space is false. As I said earlier, it's similar to the way one might show a deductive argument to be invalid: provide an interpretation in which the premises are true but the conclusion is not.

I mentioned two examples above: bugs on a sphere would not encounter an edge to their space, but their universe is finite, nevertheless. The example illustrates the distinction between infinite and unbounded, a distinction that Lucretius and others would not easily recognize until the rise of non-Euclidean geometry and modern topology. The second example is Poincaré's disk people (see Poincaré 1952). These two negative thought experiments both accept the set-up and the phenomenon of the initial Lucretius thought experiment (i.e., we never come to an edge). They deny that the conclusion (infinite space) follows from this.

John Searle and Judith Thomson have produced two of the most famous thought experiments of recent times. They were briefly described earlier. Searle's Chinese room thought experiment imagines a person in a room with an input slot and an output slot through which pass messages in Chinese writing. The person inside has a book that tells him, on a given input, what the output should be. This set-up would pass the Turing test; that is, it can think, according to the view of AI (strong artificial intelligence). But, Searle claims, obviously, the person doesn't understand Chinese at all.

There have been numerous challenges to this thought experiment, but none attack the phenomenon. No one denies that there could be such a man in a room receiving and sending messages in Chinese in accord with a book of instructions, yet not understanding the Chinese messages at all. Such an attack, were one to exist, would be a counter thought experiment, as I use the term. Instead, the challenge is usually that Searle has drawn the wrong inference from the phenomenon. One claim, for instance, is that it is not the man in the room that passes the Turing test, rather, it is the whole system: the man, the room, and the instruction book. And it is the whole system that understands Chinese, not any part of it, such as the man alone.

Thomson imagined a person hooked involuntarily to a famous violinist (who happened to be unconscious of what happened). The violinist is innocent and has a right to life. The healing process will take nine months, connected all the while; and he would die without being connected for this duration. Though it might be a very generous act to donate one's life for this period, it is surely not morally required. Abortion is analogous to this, and so abortion is morally permissible, even though (for the sake of the argument) it is granted that the fetus is an innocent person with a right to life. Thomson's thought experiment helps us to make a conceptual distinction: 'right to life' does not equal 'right to what is needed to sustain life'. The violinist/fetus has the former, but not the latter, which is why abortion is morally permissible.

This thought experiment has been repeatedly criticized and rejected, but attacks have not attempted to deny the possibility of actually finding one's self hooked to a violinist who must remain connected for nine months in order to survive. In short, the phenomenon of the thought experiment is not challenged.

In each of these cases, Lucretius, Searle, and Thomson, the challenge has not been directed against the phenomenon, but rather at some other point in the thought experiment. The phenomena in each of them has been undisputed. I turn now to the interesting cases where this is not so, that is, to cases when the phenomenon of a thought experiment has been the focus of attack.

Counter thought experiments

As I mentioned above, there are three examples of counter thought experiments that I want to discuss at length. First, Galileo denies the Aristotelian thought experiment concerning the moving earth; second, Mach denies Newton's thought experiment concerning absolute space; and third, Dennett denies Jackson's thought experiment concerning physicalism.

Galileo against Aristotle

From the time of Aristotle through the Middle Ages, there was a commonly used Aristotelian thought experiment to show the earth could not move. Suppose, on the contrary, that the earth does move. Then a dropped object would fall behind us as we move along; it would not fall straight down to our feet. But, as a matter of fact, it does fall straight down. Thus, the supposition of a moving earth must be false; the earth does not move.

Galileo put forward a counter thought experiment. Not only is it a gem in its own right, but it played a huge role in the development of physics. It established, in effect, the principle of relativity (often now called Galilean relativity). I have quoted it before and will again, such is its importance.

> Shut yourself up with some friend in the main cabin below decks on some large ship, and have with you there some flies, butterflies and other small flying animals. Have a large bowl of water with some fish in it; hang up a bottle that empties drop by drop into a wide vessel beneath it. With the ship standing still, observe carefully how the little animals fly with equal speed to all sides of the cabin. The fish swim indifferently in all directions; the drop falls into the vessel beneath; and, in throwing something to your friend, you need throw no more strongly in one direction than another, the distances being

equal; jumping with your feet together, you pass equal spaces in every direction. When you have observed all these things carefully (though there is no doubt that when the ship is standing still everything must happen in this way), have the ship proceed with any speed you like, so long as the motion is uniform and not fluctuating this way and that. You will discover not the least change in all the effects named, nor could you tell from any of them whether the ship was moving or standing still.

(*Dialogo*, 186f.)

Galileo's thought experiment denies the phenomenon of the Aristotelian thought experiment. If the earth were moving, a dropped object would land at our feet, not behind us as the initial thought experiment declared. Of course, this does not establish that the earth is indeed moving. The Aristotelian conclusion of a stationary earth might be true. But it does show that the empirical evidence we have is compatible with both a moving and a stationary earth. The Aristotelian thought experiment fails, since things would look the same regardless. This is a counter thought experiment. It does not challenge the general background premises or the chain of reasoning to the conclusion. Instead, it denies the phenomenon (that objects would fall behind) in the original thought experiment.

Mach against Newton

The background to Newton's famous thought experiment concerns rival understandings of the nature of space. Newton's absolutism (often called 'substantivalism') is the view that space is a substance that exists without depending on anything else. It is the source of inertia. Relationalism is the standard rival view: space is a system of relations. If there were no bodies, there would be no space. Leibniz, of course, is the prime representative, though he expressed his views most clearly only after Newton's thought experiment.

I hold space to be something merely relative, as time is; that I hold it to be an order of coexistences, as time is an order of successions. For space denotes, in terms of possibility, an order of things which exist at the same time, considered as existing together; without enquiring into their manner of existing. And when many things are seen together, one perceives that order of things among themselves.

(Leibniz, *Leibniz-Clarke Correspondence*, 25–6)

A 'Leibniz shift' would move the whole universe to the right, or mirror reflect it, and so on. But such a thing, Leibniz claimed, is impossible.

I say then, that if space was an absolute being, there would something happen for which it would be impossible there should be a sufficient reason. Which is against my axiom. And I prove it thus. Space is something absolutely uniform; and, without the things placed in it, one point of space does not absolutely differ in any respect whatsoever from another point of space. Now from hence it follows (supposing space to be something in itself, besides the order of bodies among themselves,) that 'tis impossible there should be a reason, why God, preserving the same situations of bodies among themselves, should have placed them in space after one certain particular manner, and not otherwise; why every thing was not placed the quite contrary way, for instance, by changing East into West. But if space is nothing else, but that order or relation; and is nothing at all without bodies, but the possibility of placing them; then those two states, the one such as it now is, the other supposed to be the quite contrary way, would not at all differ from one another. Their difference therefore is only to be found in our chimerical supposition of the reality of space in itself. But in truth the one would exactly be the same thing as the other, they being absolutely indiscernible; and consequently there is no room to enquire after a reason of the preference of the one to the other.

(*Leibniz-Clarke Correspondence*, 26)

Newton expressed his absolutism in the following much quoted passage:

Absolute space, in its own nature, without relation to anything external, remains always similar and immovable. Relative space is some movable dimension or measure of the absolute spaces; which our senses determine by its position to bodies . . .

(*Principia*, 6)

The bucket thought experiment, surely one of the most famous thought experiments ever, is described as follows:

. . . the surface of the water will at first be flat, as before the bucket began to move; but after that, the bucket by gradually communicating its motion to the water, will make it begin to revolve, and recede little by little from the centre, and ascend up the sides of the bucket, forming itself into a concave figure (as I have experienced), and the swifter the motion becomes, the higher will the water rise, till at last, performing its revolutions in the same time with the vessel, it becomes relatively at rest in it.

(*Principia*, 10)

In stage I, the surface of the water is flat and the water is at rest with respect to the bucket. In stage II, the water rotating with respect to the bucket. In stage III, the water at rest with respect to the bucket, but the surface is concave. What's the difference between I and III? Newton offers what seems like the best explanation (and possibly the only one): the water (as well as the bucket) is rotating with respect to space itself.

Of course, the bucket experiment can easily be performed as a real experiment, which presents a problem. The rest of the universe is obviously present around us, something a relationalist might appeal to. Thus, a second thought experiment is needed and is perhaps even more effective than the bucket. Newton imagines two globes in otherwise empty space.

> It is indeed a matter of great difficulty to discover . . . the true motions of particular bodies from the apparent; because the parts of that immovable space . . . by no means come under the observation of our senses. Yet the thing is not altogether desperate . . . For instance, if two globes, kept at a distance one from the other by means of a cord that connects them, were revolved around their common centre of gravity, we might, from the tension of the cord, discover the endeavour of the globes to recede from the axis of their motion . . . And thus we might find both the quantity and the determination of this circular motion, even in an immense vacuum, where there was nothing external or sensible with which the globes could be compared. But now, if in that space some remote bodies were placed that kept always position one to another, as the fixed stars do in our regions, we could not indeed determine from the relative translation of the globes among those bodies, whether the motion did belong to the globes or to the bodies. But if we observed the cord, and found that its tension was that very tension which the motions of the globes required, we might conclude the motion to be in the globes, and the bodies to be at rest . . .
>
> (*Principia*, 12)

Leibniz had no reply to this. Position and velocity are not observable, but acceleration is. The bucket and the rotating globes seemed to establish absolutism. The first serious challenge to Newton on rotation (i.e., accelerating bodies) was from Berkeley and Mach.

Mach begins his challenge to Newton with an assertion of empiricism and a new outlook on inertia. In standard Newtonian mechanics, for instance, we explain the flattening of the earth's poles and bulging of the equator in terms of the earth's rotation. And we presume that if instead of the earth rotating, the stars rotated around the earth, then the bulging of the equator would not happen. Mach takes this to be a serious mistake and that inertial forces ought to arise equally either way. This is an expression of what has come to be known as 'Mach's principle'. With this empiricist-inspired principle in the background, we come to Mach's counter thought experiment in his *Science of Mechanics*:

> Newton's experiment with the rotating water bucket simply informs us that the relative rotation of water with respect to the sides of the vessel produces *no* noticeable centrifugal forces, but that such forces *are* produced by its relative rotation with respect to the mass of the earth and the other celestial bodies. No one is competent to say how the experiment would turn out if the sides of the vessel increased in thickness and mass till they were ultimately several leagues thick.
>
> (Mach 1960, 284)

Mach's strategy is rather clear. He proposes a new theory: the source of inertia is not space, but rather is very large amounts of mass. He rejects Newton's bucket and two spheres thought experiments in the narrow sense, that is, he denies the phenomena that Newton claimed would be observed. The water in a rotating bucket with very thick walls would not climb the wall of the bucket. And in an empty universe the two balls would not act the way Newton says, but would instead move together because of the tension in the cord connecting them. Mach does not literally assert these things; I am taking a liberty. He merely remarks, somewhat rhetorically, 'who could say what would happen?' But the point is perfectly clear: the scenarios I described are as likely as Newton's. These are counter thought experiments; they deny the phenomena of the initial thought experiments.

Dennett against Jackson

Qualia are the subjective aspects of experience, feelings of hunger, pleasure, anger, and sensations of colour, smell, and so on. (One quale, many qualia.) They are accessible to introspection. The status of qualia is central to the mind-body problem. Physicalists claim that there is nothing over and above physical facts. So, qualia present some sort of challenge. Are qualia different? Can they be reduced to the physical, or perhaps eliminated? If not, then physicalism would seem to be wrong.

Frank Jackson is a long-time champion of qualia. He produced a famous thought experiment that has been much discussed for more than two decades.

Mary is a brilliant scientist who is, for whatever reason, forced to investigate the world from a black and white room via a black and white television monitor. She specialises in the neurophysiology of vision and acquires, let us suppose, all the physical information there is to obtain about what goes on when we see ripe tomatoes, or the sky, and use terms like 'red', 'blue', and so on. She discovers, for example, just which wave-length combinations from the sky stimulate the retina, and exactly how this produces via the central nervous system the contraction of the vocal chords and expulsion of air from the lungs that results in the uttering of the sentence 'The sky is blue.' (It can hardly be denied that it is in principle possible to obtain all this physical information from black and white television, otherwise the Open University would of necessity need to use colour television.)

What will happen when Mary is released from her black and white room or is given a colour television monitor? Will she learn anything or not? It seems just obvious that she will learn something about the world and our visual experience of it. But then it is inescapable that her previous knowledge was incomplete. But she had all the physical information. Ergo there is more to have than that, and Physicalism is false.

Clearly the same style of Knowledge argument could be deployed for taste, hearing, the bodily sensations and generally speaking for the various mental states which are said to have (as it is variously put) raw feels, phenomenal features or qualia. The conclusion in each case is that the qualia are left out of the physicalist story. And the polemical strength of the Knowledge argument is that it is so hard to deny the central claim that one can have all the physical information without having all the information there is to have.

(Jackson 1982, 130)

The all important knowledge argument, as it is called, runs as follows:

1. Mary knows all the physical facts about colour perception.
2. She has learned these facts having only black and while experiences.
3. When she experiences colour for the first time, she learns something new.

    Therefore, some facts about colour are not physical facts.

Before getting to Dennett's thought experiment, let me first deal with his outright rejection of any thought experiment.

Like a good thought experiment, its point is immediately evident even to the uninitiated. In fact it is a bad thought experiment, an intuition pump that actually encourages us to misunderstand its premises.

(Dennett 1991, 398)

Thought experiments depend on folk concepts; they are inherently conservative. We should expect very counter intuitive results in real science, so violating our intuitions is to be expected

(Dennett 2005, 128f.)

Dennett raises a an important point about intuitive concepts. But his in-principle dismissal of thought experiments because they use them is quite unjustified. So called folk concepts—whether they are used in thought experiments or not—can and often do lead to revolutionary results. I'll take a moment to bludgeon readers with examples.

• Galileo's thought experiment on free fall led to a new mechanics.
• Poincaré's disk thought experiment lead to very rich model of non-Euclidean geometry.
• Einstein's elevator thought experiment lead to the principle of equivalence which is central to general relativity.
• Thomson's violinist thought experiment leads to new view of the morality of abortion.

It's not just thought experiment where this happens; let me mention a few other examples:

• Arithmetic deals with very simple (folk) concepts of addition, multiplication, and division. With these we can define 'prime number' and easily prove a very profound theorem that there are infinitely many primes.
• From the very simple concepts of arithmetic we can (step by common-sense step) go on to establish the remarkable result by Gödel of the incompleteness of any set of axioms for arithmetic.
• Turing computability can readily be seen as nothing more than the elaboration of common-sense concepts of rule-governed calculation, but it leads to the unexpected result that there are uncomputable functions.

I certainly don't want to say that everything is at bottom based on folk concepts. 'Isospin', 'superego', 'magnetic field', and many other important notions are certainly not common-sense ideas at all, but must be introduced in some conjectural fashion. But it's a serious mistake to think that starting with common sense must end in common sense. Fortunately, Dennett condescends to play the thought experiment game, anyway, and he does so with considerable success.

Dennett's attack on Jackson is in the form of the following counter thought experiment. The set-up is the same as Jackson's, but the scenario is quite different.

> And so, one day, Mary's captors decided it was time for her to see colours. As a trick, they prepared a bright blue banana to present as her first colour experience ever. Mary took one look and said 'Hey! You tried to trick me! Bananas are yellow, but this one is blue!' Her captors were dumbfounded. How did she do it? 'Simple,' she replied, 'you have to remember that I know everything—absolutely everything—that could ever be known about the physical causes and effects of colour vision. So of course before you brought the banana in, I had already written down, in exquisite detail, exactly what physical impressions a yellow object or a blue object (or a green object, etc.) would make on my nervous system. So I already knew exactly what thoughts I would have (because, after all, the mere disposition to think about this or that is not one of your famous qualia, is it?). I was not in the slightest surprised by my experience of blue (what surprised me was that you would try such as second-rate trick on me). I realise that it is hard for you to imagine that I could know so much about my reactive dispositions that the way blue affected me came as no surprise. Of course it's hard for you to imagine. It's hard for anyone to imagine the consequences of someone knowing absolutely everything physical about anything!'
>
> (Dennett 1991, 399f.)

It should be quite clear at this point what is happening. Jackson's thought experiment has the following structure: there is a set-up: Mary is in a black and white room, where she learns all physical facts about perception. Next comes the phenomenon: when Mary first encounters colours, she learns something new. Finally, the result of the thought experiment, i.e., the conclusion drawn: some facts about perception are not physical, and so, physicalism is wrong. A counter thought experiment would accept the set-up, but challenge the phenomenon, which is exactly what Dennett does.

I hope the general conclusion I wish to draw from these three examples is evident: Dennett = Mach = Galileo. That is, the structure of Dennett's thought experiment is the same as Mach's and Galileo's. They are all counter thought experiments. The challenge for Dennett was not that given the thought experiment, we should resist the anti-physicalist conclusion (i.e., he is not against the broad thought experiment). Rather, Dennett's challenge is that the narrow thought experiment is faulty; the phenomenon is not as Jackson claims it would be. Mary would not learn anything new. Dennett, Mach, and Galileo each deny the phenomena of the initial thought experiments.

Alternative challenges

The Newton and Jackson thought experiments might also be challenged in the broad sense (i.e., by accepting the phenomena of the thought experiment but offering a different explanation). The challenges would not be in the form of a counter thought experiment, possibly not in the form of a thought experiment at all. I'll briefly mention two examples, just for the sake of comparison.

Contra Newton, Larry Sklar introduced his notion of *absolute acceleration*. (Sklar 1976) An object or system of objects, such as the two spheres connected by a cord, might have this property. When it does, there will, for instance, be a tension in the cord joining the two spheres. They are not rotating with respect to anything, they simply have this property of absolute acceleration. It's quite bizarre, but if one thinks of quantum mechanical spin, then one gets the idea. The spin of an electron is 'intrinsic'; it cannot be transformed away in any coordinate frame. Sklar's account, unlike Mach's, does not challenge the phenomenon of Newton's thought experiment; it offers a different explanation.

Contra Jackson, David Lewis proposed the 'ability hypothesis' (Lewis 1983, 1988). It is related to the distinction between knowing how, not knowing that. One might know absolutely every fact about a bicycle, yet not know how to ride. If one learns how to ride, one is not learning a new fact or acquiring new propositional knowledge, but rather one is acquiring a skill, a new ability. When Mary leaves the laboratory and experiences red things for the first time, she is similarly learning a skill, not learning new fact. Lewis's account, unlike Dennett's, does not deny the phenomenon of Jackson's thought experiment, but rather undermines Jackson's knowledge argument by interpreting the phenomenon of the thought experiment in a different way.

Evaluation in these cases takes the form: Who offers the best explanation or interpretation of the phenomena in the thought experiment? Sklar and Lewis do not deny the phenomena of Newton's and Jackson's thought experiments. Rather they challenge the inference drawn after accepting the phenomena.

When does a counter thought experiment work?

The main point of this chapter is the modest one of pointing out the existence of a distinct class of thought experiments, namely, counter thought experiments. But once we accept their existence and get some idea of how they work, the inevitable questions to ask are: When do they work? And when do they fail? What follows is but a superficial start on addressing these questions.

Clearly, a counter thought experiment will work only when it can plausibly deny the phenomenon of the original thought experiment. I don't think anyone could reasonably hope to deny the phenomena in, say, Searle's Chinese room thought experiment. Everyone is ready to allow that a person could be in a room with Chinese characters taken in that are compared by the person inside with those in a book that tell her which Chinese characters to put out. Challenges to Searle's thought experiment have all been aimed at the inference that he drew from that phenomenon. Similarly, there is no point in rejecting the phenomenon in Lucretius' infinite space thought experiment, since it merely involves throwing a spear. To do so would involve a degree of scepticism that goes well beyond the case at hand. The same could be said of Thomson's violinist. Of course, we could wake up with an unconscious and very ill violinist hooked up to ourselves such that he will be cured if and only if he remains connected for nine months. There could be no plausibility to denying that such things could happen. Challenges to these thought experiments must be aimed at the various conclusions the thought experimenter draws from the thought experiments.

This should be uncontentious, but there are those who would disagree. Peter Geach, for instance, takes moral rules to be divine commands and he holds that God would not allow genuine moral dilemmas to exist, since we would then have to choose between different commands. Geach imagines someone saying: 'But suppose circumstances are such that observance of one Divine law, say the law against lying, involves breach of some other absolute Divine prohibition'. Geach then replies:

> If God is rational, he does not command the impossible; if God governs all events by his providence, he can see to it that circumstances in which a man is inculpably faced by a choice between forbidden acts do not occur. Of course such circumstances (with the clause 'and there is no way out' written into their description) are consistently describable; but God's providence could ensure that they do not in fact arise. Contrary to what unbelievers often say, belief in the existence of God does make a difference to what one expects to happen.
>
> (Geach 1969, 128)

The upshot, according to Geach, is that perfectly consistent thought experiments might still be illegitimate and the phenomenon not exist, because God would not allow it to happen. I mention this outlook in passing to further illustrate the range of possible opinions on thought experiments. It is not an opinion we should seriously consider.

Here are some things that seem to matter when evaluating a counter thought experiment. They are all rather obvious.

- How reliable is the initial thought experiment in the narrow sense (i.e., would the phenomenon occur)?
- How strong is the assumed background to the thought experiment?
- How similar is the phenomenon of the thought experiment to things we know and trust?
- How plausible is the phenomenon of the counter thought experiment?
- How absolutely plausible?
- How relatively plausible?

The last of these is probably the key question, but let's flesh them all out a bit by considering our three examples.

In making the case for the Aristotleans, we might note what happens when we throw some litter out the car window—it falls far behind as we move along. (I hope introducing a car is a harmless anachronism. Littering was, is, and always will be wicked.) The case for Galileo could be based on our experience of tossing things around in a car, aeroplane, or any other moving vehicle. Motion seems to have no effect. By analogy, if the whole earth were moving, our experience should be as in a car, or aeroplane, and so on.

In making the case for Newton, we might note that we have often seen water climb rotating buckets and we have felt the tension in a string holding a rock that is spinning around us. The two globes thought experiment assumes they would act the same in empty space and this seems very plausible. Mach, on the other hand, proposes a new theory: mass is the cause of inertial motion. There is no empirical evidence for this; it's motivated by his rather strict empiricism. (Remember the fate of Mach's empiricist-inspired anti-atomism.) However, it would seem that acceleration should be on par with position and velocity—relative, otherwise not detectable.

Given that Mach's account is possible, he undermines to some extent our degree of belief we had in the phenomenon of Newton's thought experiment, i.e., that the cord's tension would be maintained. But Mach's counter thought experiment is certainly not as plausible as Newton's, since only the latter replicates our normal experience. Consequently, it is a weak attack on Newton's thought experiment, and hence, a weak attack on absolute space.

The case for Jackson might begin by noting that in general, mental things don't seem like physical things, and more specifically, when people, for instance, acquire eyesight late in life, they appear to learn something new. Mary, the colour scientist, seems like an extreme case of this; hence, Jackson's narrative appears initially plausible. But we don't know anyone who knows everything even about some triviality, much less all the physical facts. So, the analogy with things we already know is very weak. There is a superficial similarity to Plato's cave, but with vastly greater assumptions.

The case for Dennett might begin with noting that for various philosophical reasons, physicalism seems right (i.e., problems with dualism, etc.). And, to repeat, we have no idea what it would be like for someone to know all the physical facts. As a story, Dennett's narrative about Mary seems coherent and intelligible. It would appear then that Dennett's counter thought experiment is just as plausible as Jackson's, even though neither is in itself very plausible. Dennett is quite aware of this:

> My variant was intended to bring out the fact that, absent any persuasive argument that this could not be the way Mary would respond, my telling of the tale had the same status as Jackson's: two little fantasies pulling in opposite directions, neither with any demonstrated authority.
>
> (2005, 105)

That's enough. Jackson is neutralized, if not refuted, and consequently, the Mary thought experiment is a failure.

Comparatively, I would say that Galileo is completely triumphant; the Aristotelian thought experiment is destroyed. Newton is slightly weakened but not seriously damaged; Mach is an alternative, but it is nowhere near as plausible. Jackson is nullified, since Dennett's alternative story is equally plausible. It's a tie, as far as the thought experiments go, which probably leaves Dennett the victor. As is sometimes the case in war, one side can win by simply not losing.

These evaluations are, of course, quite rough and doubtless open to objection. They are only preliminary and should not be taken too seriously. They merely illustrate the kinds of consideration involved. My main aim is to determine how counter thought experiments work in general, not to evaluate particular instances. I do, however, wish to explore the comparative nature of the thought experiment-counter thought experiment pair by briefly examining a simple proposal.

A ratio test

For quite some time, it has been common to think of the evaluation of scientific theories as taking place comparatively. Kuhn's paradigms and Lakatos's research programmes are evaluated (at least in part), by comparing them with rivals. Much the same can be said of counter thought experiments. There are, however, important differences. Comparative theory evaluation is usually over the long haul and it is the whole theory/paradigm/programme that is being compared. By contrast, thought experiments and counter thought experiments go head to head and the evaluation is direct and immediate.

In trying to capture this, we might try the following *Ratio Test*, as I shall call it. Assign a probability to the phenomenon of a thought experiment, given the thought experiment set-up. I should readily admit and even stress that this not intended to be realistic; I doubt these things can be quantified. But it might shed a little light on the structure of counter thought experiments, nevertheless.

Let *i-phen* = phenomenon in the initial thought experiment (e.g., action of Newton's two spheres in empty space, or Mary learning something new when she leaves the laboratory), and *c-phen* = phenomenon in the counter thought experiment (e.g., action of the two spheres according to Mach, or the actions of Mary in Dennett's thought experiment). Assign probabilities to these, e.g., Prob(Mary learns something new) = r.

It would seem that a counter thought experiment is successful, if Prob(c-phen)/Prob(i-phen) > 1, and is not successful, if Prob(c-phen)/Prob(i-phen) << 1.

Why does the second claim not use ≤, which would be a simple denial of the first? The reason has to do with a complication I have not mentioned, but will on reflection be obvious.

Presenting a counter thought experiment is perhaps like the defence presenting an alternative account of the facts in a legal case. The prosecution must make its case 'beyond a reasonable doubt'. The defence need not match that high standard, but need only make a case for a slightly plausible alternative—win by not losing. Even if we think the prosecution's story is more likely, the possibility presented by the defence is enough to undermine our initial confidence. In probabilistic terms, the defence can do its job successfully even when its case has a probability well below ½, just as long as the probability isn't too low. Mach's counter thought experiment might plausibly fall in this range. It is not plausible in its own right, but it could be just plausible enough to undermine our initial assessment of Newton's thought experiment.

In general, the range of plausibility of counter thought experiments is great. Some counter thought experiments might be highly compelling in their own right, as was Galileo's. Others might be weak in their own right, but still strong enough to cast doubt on the rival account, as Mach's perhaps did to Newton's.

There are also cases where the ratio test might break down badly. This will happen in "how possible" thought experiments. These are thought experiments that don't try to establish a result concerning how things are, but only try to show how something is possible. Darwin provided examples in discussion of the evolution of particular characteristics that seemed problematic. How could the eye evolve or the giraffe acquire a long neck? Darwin's thought experiment would show a possible evolutionary route. It was not intended to be true, only to show that the particular biological characteristic is not a counter example to the theory of evolution. As long as its probability is not equal to zero, the thought experiment is a success. One of Darwin's foremost early critics, Fleming Jenkins, produced counter thought experiments (involving 'blended inheritance') that aimed to show the evolutionary account Darwin provided is not possible. In other words, he constructed counter thought experiments with probability virtually equal to one. And that is what it had to be, since he needed to show that the phenomenon in Darwin's was not even possible. (See Lennox 1991 for an account of Darwin and Jenkins.)

The ratio test is quite inappropriate in cases such as these, since almost inevitably $\text{Prob(c-phen)}/\text{Prob(i-phen)} \gg 1$. This will happen even when the counter thought experiment is only moderately plausible, since the initial thought experiment is only meant to show possibility, not likelihood. The ratio test is at best a first stab; it is not adequate as it stands. Counter thought experiments that aim to undermine 'how possible' thought experiments will have to be evaluated some other way.

Is it possible to give a general account—perhaps quite different from the one I have sketched—of what makes a counter thought experiment an effective one? This is wholly unexplored territory, but definitely worth further investigation, as are all areas of the remarkable topic of thought experiments.

4   Mathematical thinking

Platonism, according to one unsympathetic commentator, 'assimilates mathematical enquiry to the investigations of the astronomer: mathematical structures, like galaxies, exist independently of us, in a realm of reality which we do not inhabit but which those of us who have the skill are capable of observing and reporting on' (Dummett 1973, 229). I'll call this view of mathematics 'π in the sky'—but not scornfully, though perhaps with a touch of self-mockery, since I think it is true. Mathematics is best accounted for by appeal to real Platonic entities; not only do they provide the grounds for mathematical truth, but these abstract objects are also somehow or other responsible for our mathematical intuitions and insights.

This chapter is really stage-setting for the next. My aim is to liken thought experiments to mathematical thinking. But, in order to do that I need first to establish what I take to be the real nature of mathematical thought itself; hence this defence of Platonism. I will take the analogy even further toward the end of this chapter, where visual reasoning in mathematics will be tied to thought experiment.

Combinatorics

Paul Benacerraf, in a very influential paper (1973), contrasts the semantically motivated Platonistic account with the 'combinatorial' view, as he calls it. Hilbert's formalism is a paradigm combinatorial account of mathematics, but Benacerraf intends to include more under this umbrella term.

> combinatorial accounts . . . usually arise from a sensitivity to [epistemological problems]. . . . Their virtue lies in providing an account of mathematical propositions based on procedures we follow in justifying truth claims in mathematics: namely, proof.
>
> (1973, 409)

Benacerraf is as unhappy with combinatorial views of mathematics as he is with semantical accounts. Their virtues come at a price. They make sense of *knowledge,* he thinks, but only at the cost of *not* being able to make sense of truth and reference. In other words, combinatorial and Platonist accounts have opposite virtues and opposite vices. What I first want to cast doubt on is the belief that combinatorial views have any virtues at all.

'Proof' is taken by Benacerraf to mean formal derivability. That is, to prove a theorem is really to prove that

$$\vdash \text{Axioms} \supset \text{Theorem}$$

is a logical truth.

Such a logical relation may well exist between the first principles of a mathematical theory and any of its theorems, but what role does this logical relation play in bringing about *knowledge*? Can this relation be used to explain epistemic facts? Establishing such a logical relation is usually a non-trivial task. Philosophers and mathematicians who champion such a view of mathematical proof usually omit mentioning the enormous difficulties. Or when they do admit, as Benacerraf does, that 'this is of course an enormous task', the concession is down-played by appearing only in a footnote (1973, 409n). To get a feel for the enormous difficulties that can be involved consider the following simple example.[1]

A chess-board has $8 \times 8 = 64$ squares. Remove two opposite corner squares, leaving 62. A domino covers two adjacent squares. Question: Is it possible to tile, that is, to completely cover the board with 31 (non-overlapping) dominoes?

The answer is, 'No'. Here is a proof: Each domino must cover one black and one white square. Thus, any tiling could only be done if the number of black and white squares is equal, which is not the case here, since there are two fewer white squares than black ones. Thus, tiling the board is impossible.

Of course, this proof is in no way combinatorial; nor is it a formal derivation. The theorem, however, actually represents a logical truth. To see this we first need to translate the problem into logical terms, which we can do as follows. Each square gets a name which represents the proposition that it is covered.

The proposition 'p&q' means 'Square p is covered and square q is covered'. Any domino covering p will also cover a square beside p; moreover, if it is covering p and q then it is not covering p and r, etc. Thus we have:

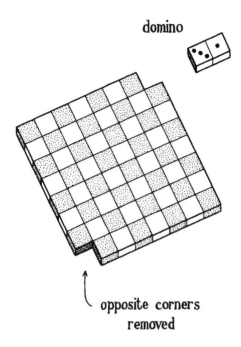

domino

**opposite corners
removed**

Figure 19

[(p&q)v(p&r)v(p&s)v(p&t)
& [[(p&q) ⊃ –((p&r)v(p&s)v(p&t))]
& [(p&r) ⊃ –((p&q)v(p&s)v(p&t))]
& [(p&s) ⊃ –((p&q)v(p&r)v(p&t))]
& [(p&t) ⊃ –((p&q)v(p&r)v(p&s))]]]

This describes the situation for *one* square P. Continuing on in this way to describe the whole board we would have a sentence about sixty-two times as long as this. Such a giant proposition is, as a matter of fact, a contradiction; so its negation is derivable in any complete system of propositional logic. But who could derive it? And who thinks his or her knowledge of the tiling theorem is based on this logical truth?

Is this the only way to express the problem in logical terms? Undoubtedly, no. Perhaps there are quantificational versions which are more manageable. But whatever the translation for combinatorial purposes is, notice how much worse the situation becomes for the would-be combinatorialist by enlarging the size of the board. The

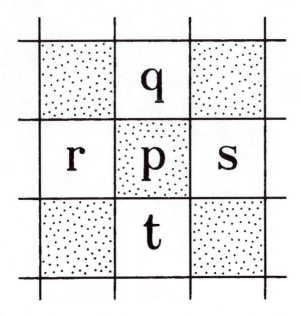

Figure 20

short proof works for any n×n chess-board (with two opposite corners removed). It establishes the result with as much ease for a board with $10^{100} \times 10^{100}$ squares as the standard 8×8 case. But combinatorial proofs would have to be changed for each *n,* and they get more and more unwieldy as *n* increases.

Poincaré, in his essay 'Intuition and Logic in Mathematics' (1958), argues that logic alone cannot give us the great wealth of mathematical results. Though he holds that intuition is the main source of mathematical knowledge, Poincaré nevertheless seems willing to grant that logic can justify mathematical theorems after the fact. (See the section on discovery *vs* justification in the next chapter.) If anything, Poincaré is perhaps being too generous; logic is neither the source of the ideas to start with, nor, in any *practical* sense, is it the justification. (At least, not in a wide variety of cases, though it is certainly the justification in many.)

The epistemological consequence of this is straightforward. If mathematical knowledge is based on proof, and proof is formal derivation, then in a vast number of cases, such as the tiling example, mathematical knowledge is impossible. Obviously, this is just absurd.

Benacerraf is right in thinking any combinatorial account has trouble with truth, but wrong to think it can do full justice to our epistemological concerns. It can't do either; it is entirely without virtue. Given Benacerraf's dichotomy, this brings us back to Platonism.

What is Platonism?

In a much cited passage, the most famous of recent Platonists, Kurt Gödel,[2] remarks that 'classes and concepts may . . . be conceived as real objects . . . existing independently of our definitions and constructions'. He draws an analogy between mathematics and physics:

> the assumption of such objects is quite as legitimate as the assumption of physical bodies and there is quite as much reason to believe in their existence. They are in the same sense necessary to obtain a satisfactory system of mathematics as physical bodies are necessary for a satisfactory theory of our sense perceptions . . .
>
> (1944, 456f.)

And in the same vein:

> despite their remoteness from sense experience, we do have something like a perception also of the objects of set theory, as is seen from the fact that the axioms force themselves upon us as being true. I don't see any reason why we should have any less confidence in this kind of perception, i.e., in mathematical intuition, than in sense perception.
>
> (1947, 484)

In making their position clear, Platonists are often forced to resort to analogies and metaphors. Gödel's is the most famous, but my favourite is one by the mathematician G.H. Hardy, which I'll quote at length since it is not as well known as it should be.

> I have myself always thought of a mathematician as in the first instance an *observer,* a man who gazes at a distant range of mountains and notes down his observations. His object is simply to distinguish clearly and notify to others as many different peaks as he can. There are some peaks which he can distinguish easily, while others are less clear. He sees A sharply, while of B he can obtain only transitory glimpses. At last he makes out a ridge which leads from A, and following it to its end he discovers that it culminates in

B. B is now fixed in his vision, and from this point he can proceed to further discoveries. In other cases perhaps he can distinguish a ridge which vanishes in the distance, and conjectures that it leads to a peak in the clouds or below the horizon. But when he sees a peak he believes that it is there simply because he sees it. If he wishes someone else to see it, he *points to it,* either directly or through the chain of summits which led him to recognize it himself. When his pupil also sees it, the research, the argument, the *proof* is finished. The analogy is a rough one, but I am sure that it is not altogether misleading. If we were to push it to its extreme we should be led to a rather paradoxical conclusion; that there is, strictly, no such thing as mathematical proof; that we can, in the last analysis, do nothing but *point;* that proofs are what Littlewood and I call *gas,* rhetorical flourishes designed to affect psychology, pictures on the board in the lecture, devices to stimulate the imagination of pupils.

(1929, 18)

Platonism involves several ingredients:

(I) Mathematical objects exist independently of us, just as do physical objects. As Dummett puts it: 'Mathematical statements are true or false independently of our knowledge of their truth-values: they are rendered true or false by how things are in the mathematical realm' (1967, 202). Or as Hardy remarks: 'In *some* sense, mathematical truth is part of objective reality. . . . When we know a mathematical theorem, there is something, some object which we know' (1929, 4).

(II) Mathematical objects are abstract; they exist outside space and time. This distinguishes Platonism from non-Platonic forms of mathematical realism such as the empiricism of Mill or Kitcher (1983), or the account of Irvine (1986) which uses Armstrong's universals (which depend on the physical world for their existence).

(III) We learn about mathematical objects as a result of the mind's ability to somehow grasp (at least some of) them. This is where analogies have played the biggest role. Thus, Gödel likens this grasping to the 'perception' of physical objects and Hardy to 'seeing' a peak in a distant mountain range and 'pointing' to it. On the other hand, we needn't directly perceive all the mathematical objects that we know about; some might be known through conjecture, as theoretical entities are known in physics. For instance, we directly see grass and the moon, but not electrons; we know about elementary particles since they are part of the scientific theory which is (we hope) the best explanation for the white streaks in cloud chambers that we do see.[3] Similarly, some mathematical objects and some mathematical axioms are conjectured, not directly intuited.

(IV) Though it is *a priori* (i.e., independent of the physical senses), the mathematical learning process is not infallible.[4] We don't deny the existence of teacups even though we sometimes make perceptual mistakes about them. And since much of our mathematical theorizing is conjectural, it is bound to be wrong on occasion. Thus, Gödel rightly remarks: The set-theoretical paradoxes are hardly more troublesome for mathematics than deceptions of the senses are for physics' (1947, 484).

Of these four ingredients the first two have to do with the metaphysics of Platonism, the latter two with its epistemology. The third seems to be the most problematic; at least it has been the subject of most of the attention and ridicule from physicalists. But before fending off those attacks, let's quickly go over the reason for being sympathetic with Platonism in the first place. The π in the sky account of mathematics does, after all, have numerous virtues.

(A) It makes 'truth' in mathematics well understood. A mathematical sentence is true or false in just the same way that a statement about an everyday physical object is true or false. Conventionalists and constructivists, on the other hand, have to do a great deal of fancy footwork to explain what is meant by saying that '2 + 2 = 4' is true. Classical logic with all its power is a legitimate tool for mathematical employment. Thus we can continue in the use of our traditional mathematical techniques (such as indirect proofs); practice will not be hampered and does not have to be re-interpreted somehow.

(B) Platonism explains our intuitions, our psychological sense that various theorems really are true and must be so. No conventionalist or empiricist account can do justice to these psychological facts. Quine, for example, accepts a watered-down Platonism which includes (I), (II), and (IV) but denies (III); that is, he denies we can perceive abstract objects; instead he claims we learn mathematical truths by testing their empirical consequences. But this leaves it an utter mystery why '3 > 2' seems intuitively obvious while 'protons are heavier than electrons' does not.

(C) Finally, the Platonistic picture gives a more united view of science and mathematics. First, they are methodologically similar (though not identical) kinds of investigation into the nature of things. Second, science also needs abstract entities; the best account of laws of nature takes laws to be relations among universals which are abstract entities (see chapter five, which follows Armstrong 1983, Dretske 1977, and Tooley 1977). The case for this account of laws and the case for mathematical Platonism are similar.

This virtue must, however, be sharply distinguished from a merely alleged virtue that science needs mathematics. Quine and Putnam, for example, have made 'indispensability arguments' to the effect that in order for science to work, mathematics must be true. This is the least convincing reason for Platonism; the case for it is best made from pure mathematics alone. If Hartry Field's *Science Without Numbers* works against these indispensability arguments, then it has merely undermined the weakest argument for Platonism; the others are left intact.[5]

Of course, there is always one more virtue that any theory might have. It is comparative, having to do with the relative merits of its rivals. Here is not the place to list the many failings of formalism and other types of conventionalism, nor of intuitionism and other forms of constructivism, nor of physicalism in any of its versions. Readers who find the ontological richness of Platonism distasteful should simply recall that the alternatives are even less palatable.

Platonism, however, is not without its problems. An oftheard complaint is that it is completely redundant. What possible difference would it make if there were no Platonic objects? Wouldn't everything else still be the same? Moreover, analogies often cut two ways. Gödel says there is a kind of perception of abstract objects which is similar to ordinary sense perception. The attractiveness of the π in the sky view rests in part on this analogy. But just what sort of perception is it? Critics claim that the account will never get past the hand-waving metaphors. For example, it has been called 'flabby' by Charles Chihara; and one of the most influential critics, Paul Benacerraf, remarks that the Platonist 'will depict truth conditions in terms of conditions on objects whose nature, as normally conceived, places them beyond the reach of the better understood means of human cognition (e.g., sense perception and the like)' (1973, 409). Benacerraf suggests it is mysterious because we know nothing at all about the kind of perception of sets that Gödel mentions, and it may actually be impossible, since abstract objects if they did exist would be unknowable. The argument for this last objection is indeed plausible: to know about anything requires a causal connection between the knower to the known, and there can be no such connection between things inside space and time (i.e., us) and things outside space and time (i.e., sets and other abstract objects).

In sum, the cost of having a Platonic explanation for our intuitive beliefs, as well as having a nice neat semantics and all the other virtues I mentioned, is very high: the explanation is at best flabby, our ontology has a vast store of redundant elements, and our knowledge is at best highly mysterious, and perhaps even impossible.

Even some of the friends of abstract objects such as W.V. Quine and Penelope Maddy shrink away from the epistemological aspects of Platonism, refusing to believe that we can perceive sets in the full π in the sky sense. All of these objections to Platonism stem from a common motivation: empiricism, physicalism, and naturalism. The redundancy objection says we can do science without abstract objects; Chihara's flabbiness objection says the Platonic explanation is not the same as explanations in physics; the mysteriousness objection says we do understand physical perception, but not the perception of abstract objects; the causal objection says knowledge requires a physical connection between object and knower. Faint-hearted Platonists like Quine and Maddy try to reconcile their realistic views with these physicalistic objections rather than trying to meet the objections head on.

My strategy in the balance of this chapter is, first, to attack the faint of heart, then to take up the cudgels against the physicalistically inspired objections to fully-fledged π in the sky.

The faint of heart

Willard Quine and Penelope Maddy are fellow travellers of the Platonist. They are both realists about sets: sets exist and are the truth makers for mathematical sentences. Where they back off accepting π in the sky is over epistemology; at this point physicalist sentiment prevails (though quite differently) in each case. Mathematics, in Quine's well-known metaphor, is part of our web of belief (1960). It is in the centre, and hence relatively immune from revision, but it interacts with all the other elements. It is very much like theoretical science, which is tested in hypothetical fashion via observational consequences. Schematically:

Mathematics statements
Physics statements
_____

∴ Observation statements

True observation sentences count *for* the mathematics and physics used; false observation sentences count *against*. The physicalist's epistemological scruples are largely satisfied; there are no claims here about seeing sets any more than there are claims about seeing electrons. All that is perceived are streaks in cloud chambers and the like which serve to support not just quantum mechanics but the theory of differential equations as well.

There are two problems which upset this view. First, as Parsons remarks, 'it leaves unaccounted for precisely the *obviousness* of elementary mathematics' (1980, 151). There are no sentences of quantum mechanics, or of theoretical genetics, or of theoretical psychology, etc. which feel obvious or seem as if they have to be true. Yet such sentences abound in mathematics. No matter what Quine says, our conviction that $2+2=4$ does not stem from laboratory observations, no matter how carefully performed or often repeated.

Second, Quine's account of mathematics does not square with the history of science. Mathematics certainly does have a history; it is naive to think that a mathematical result, once established, is never overturned. (Lakatos 1976 is a good antidote to thinking otherwise.) But the sciences have had nothing to do with this. It is not that mathematics and physics don't interact; obviously they do. The discovery of non-Euclidean geometries made general relativity thinkable; and the success of general relativity stimulated a great deal of further work on differential geometry. But their interaction is more psychological than logical. In the entire history of science the arrow of *modus tollens* (after an unexpected empirical outcome) has never been directed at the heart of mathematics; it has always been a theory with physical content which has had to pay the price.

Maddy protests that Quine's account is at odds with mathematical practice. *Contra* Quine, she rightly notes that '(i) in justifying their claims, mathematicians do not appeal to applications, so [Quine's] position is untrue to mathematical practice, and (ii) some parts of mathematics (even some axioms) aren't used in applications, so [Quine's] position would demand reform of existent mathematics' (1984, 51).

The way mathematics is applied to science is not in the form of additional premises added to physical first principles, but rather as providing models. A scientist will conjecture that the world (or some part of it) W is isomorphic to some mathematical structure S. Explanations and predictions are then made by computing within S and translating back to the scientific language. If it is empirically a failure, no one would or should dream of modifying S; rather one would look for a different structure S' and claim that W is isomorphic with it. Energy, for example, was modelled on the real numbers; but the 'ultraviolet catastrophe' (of black-body radiation) which led to the quantum theory was no threat to the theory of real numbers; instead, the old conjectured isomorphism was dropped and energy is now modelled on the integers (i.e., the energy operator has a discrete spectrum).

For any way that the world W could be, there is some mathematical structure S which is isomorphic to W. I suspect that this fact about applied mathematics is what undermines Quine, since nothing that happens in the world would change our views about *S* itself. Experience could only change our view that the physical world is like *S'* rather than like *S*. The corollary is that we cannot learn about the properties of these various mathematical structures by examining the physical world; before we can conjecture that *W* is like *S* we must know about *S* independently.

Maddy may be sympathetic with some of this, but her physicalistic sentiments incline her away from a fully-fledged Platonism. Like Quine, though, she comes very close to π in the sky without actually embracing it.

When Maddy looked into her refrigerator she saw three eggs; she also claims to have seen a set. The objection that sets aren't anywhere in space or time, much less in her refrigerator, is answered: 'It seems perfectly reasonable to suppose that such sets have location in time—for example, that the singleton containing a given object comes into and goes out of existence with that object. In the same way', she continues, 'a set of physical objects has spatial location in so far as its elements do. The set of eggs, then, is located in the egg carton—that is, exactly where the physical aggregate made up of the eggs is located' (1980, 179). Her belief that there is a three-membered set of eggs in her refrigerator is, according to Maddy, a perceptual belief. But it doesn't seem plausible to suggest that seeing the threeness of the set is like seeing the whiteness of the eggs. In order for it to be a perceptual rather than an inferential belief, she would also have to see the one-one, onto function between the set of eggs and the ordinal number three, which means she would have to keep the number three in her refrigerator, too.[6]

There are other oddities with Maddy's sets. Consider the fact that the 'existence' of Santa Claus is parasitic on the real existence of red things, jolly things, white-bearded things, etc. which allows Santa to be defined as the jolly, white-bearded fellow dressed in red. As we shall see this can lead to trouble. Maddy thinks sets are genuine natural kinds, at least the ones which are sets of physical objects. 'This kind can be dubbed by picking out samples [e.g., the set of eggs in her refrigerator]. Particular sets and less inclusive kinds can then be picked out by description, for example, "the set with no elements" for the empty set, or "the sets whose transitive closures contain no physical objects" for the kind of pure sets' (1980, 184).

This means the empty set exists only courtesy of the existence of sets of physical objects. Set theory in its completely pure form, independent of any urelements, can't exist. In its pure form, we start with ∅, then reiterating the set-theoretic operations build up the hierarchy: ∅, {∅},

{Ø, {Ø}}, etc. But Maddy can have none of this; there can be no such thing on her account as genuinely pure sets which aren't somehow or other dependent on physical objects. There can only be a set theory based on the impure kind, those which have urelements, real physical entities, as members. Of course, Maddy can define pure sets out of impure ones and thereby have a pure set theory, but this requires the existence of impure sets to start with. Pure sets are parasitic on impure sets; they can only be defined into existence in terms of them.

However, we may object, what if there were no physical objects at all? Wouldn't it still be true that {Ø} < {Ø, {Ø}} and that $2 + 2 = 4$? Leibniz thought that God freely chose to create the physical world, but that the laws of logic and mathematics are true independent of God. While I don't endorse this (or any) theology, the sentiment which motivates Leibniz seems right: there is something contingent about the existence of the physical world, a contingency not shared by the laws of mathematics. Two plus two does equal four whether there are any apples to instantiate this fact or not, and the theorems of set theory are true whether or not there are any urelements. Sets of physical objects depend on the prior existence of pure sets, not vice versa.

A flabby explanation?

Charles Chihara is an unsympathetic critic of Platonism. He finds it a 'flabby' account and thinks that 'it is at least as promising to look for a naturalistic explanation [of the commonality of mathematical experiences] based on the operations and structure of the internal systems of human beings' (1982, 218). His reasons for being dismissive, however, are not persuasive. Chihara heaps scorn on the Platonists' analogy between seeing physical objects and seeing mathematical objects. As an example, he considers the similarities (or rather dissimilarities, in his view) between 'believing in sets' and 'believing in molecules'. He then notes that we have the results of careful experiments, novel predictions, and so on in the case of molecules, but we have nothing like this in the case of sets. He rhetorically asks, 'What empirical scientist would be impressed by an explanation this flabby?' (1982, 217).

Unfortunately, Chihara has completely misunderstood the structure of the analogy.[7] Let me clarify by first distinguishing the grounds of our knowledge in physics. There are the everyday objects of perception, such as tables, chairs, trees, the moon, etc. Realism about these objects is contrasted with phenomenalism. Unlike Berkeley, the vast majority of us are realists about observable objects. Why? The usual reason is that the reality of those objects is the best explanation for our common perceptual experiences. On the other hand there are the theoretical entities of science such as electrons, magnetic fields, genes, etc. We accept theories about these based on their ability to explain, to systematize, and to predict the behaviour of observable objects. Realists in this domain (usually called 'scientific realists') are contrasted with instrumentalists; the former say that reasons to adopt a theory are reasons to believe that it is true and hence reasons to believe that the theoretical entities involved exist, while instrumentalists claim that acceptable theories are to be understood only as being empirically adequate.

So how then does the Platonist's analogy with this work? Gödel was quite explicit in dividing our set-theoretic knowledge into two sources. Where he claimed that we had intuitions, or 'something like a perception . . . of the objects of set theory, as is seen from the fact that the axioms force themselves upon us as being true' (1947, 484) he meant in the elementary parts, i.e., concerning things we would call *obvious*. This would correspond to the physical perception of such objects as tables, chairs, apples, the moon, and white streaks in cloud chambers. He also held that some of our knowledge is conjectural; we hypothesize an axiom and subsequently come to believe our conjecture because we believe its consequences, which are directly obvious. For example, Gödel remarks,

> There might exist axioms so abundant in their verifiable consequences, shedding so much light upon a whole field, and yielding such powerful methods for solving problems . . . that, no matter whether or not they are intrinsically necessary, they would have to be accepted at least in the same sense as any well-established physical theory.
>
> (1947, 477)

The boundary between those sets which can be perceived and those which cannot is undoubtedly fuzzy, just as in the case of physical objects. Moreover, things can be made even more complicated by introducing the notion of theory-laden perceptions, which are no doubt as prevalent in mathematics as in physics. But these concerns are not the central issue. The crucial point is that the analogy Platonists think exists between mathematical perception and physical perception has to do with the perception of macro-objects, not molecules.

The canons of evidence in the two cases are different. We believe in tables because we have table experience. On the other hand, we believe in molecules, not because we have experience of molecules, but because we subject the molecule theory to a wide battery of tests—we demand explanatory power, novel predictions, etc. Similarly, axioms of large cardinals, or that V=L, or the Continuum Hypothesis have been and continue to be thoroughly scrutinized; and in so far as they measure up (i.e., have plausible consequences), we accept them as being true. Gödel simply wants us to believe in the existence of some sets, the ones we can perceive, for the same sorts of reason that we reject Berkeley and happily believe in the existence of macroscopic physical objects.

Finally, Chihara castigates Platonism for being so much cabbalistic claptrap.

> Surely, there is something suspicious about an argument for the existence of sets that rests upon data of so unspecified and vague a nature, where even the most elementary sorts of controls and tests have not been run. It is like appealing to experiences vaguely described as 'mystical experiences' to justify belief in the existence of God.
>
> (1982, 215)

Not so. Anyone at any time can have the experience of green grass or of two plus two equalling four, which makes these two examples alike and yet makes both quite unlike mystical experiences which cannot be reproduced even by those who claim to have had one. Chihara is right to dismiss mystical experiences, but there is not a shred of similarity between them and mathematical experiences.

Does it matter?

One of the most common objections to Platonism—more often made in conversation than in print—is that the existence of abstract objects is irrelevant. Would things be any different if abstract objects did not exist? The question is usually asked rhetorically, the presumption being that the obvious answer says things would be exactly the same whether abstract entities did or did not exist.

But this is the wrong answer. Things would be very different. If there were no abstract objects, then we wouldn't have intuitions concerning them; '2 + 2 = 4' would not seem intuitively obvious. It is the same with teacups; if they did not exist I wouldn't see any and there would be a great mess on the table every time I tip the pot.

Of course, the question could be intended in a deeper way, as asking: How would things be different if there were no abstract objects but everything else, including our 'intuitions', remained the same? I have no answer to this, except to point out that it is just like a Berkeleian sceptic asking the same question about material objects. The situation is no more embarrassing to the Platonist than belief in ordinary material objects is to everyone else. We believe in independently existing material objects as somehow or other the cause of the phenomena; similarly, we should believe in Platonic objects as somehow or other the cause of intuitions. However, this is not the place for an attack on general scepticism, so I won't pursue the issue.

The modern brand of mathematical Platonism isn't the same as Plato's, but it has much in common with the earlier theory, and it has to face many of the same objections. The challenge that 'it doesn't matter whether abstract objects exist or not' is one of the earliest. Aristotle, for example, claims (*Metaphysics* 990A 34ff.) that Plato's forms are an unnecessary duplication of the physical world. But what Aristotle fails to consider is that the forms do so much more than merely account for the physical world; they are also the source of *our knowledge* of the physical world and the basis of moral value. Similarly, mathematical objects aren't merely the truth makers of mathematical sentences, they are somehow or other responsible for our mathematical beliefs as well.

A physicalist might concede this point only to pounce on another. Perhaps the existence of abstract objects would make a difference to our mental states, but how? What does it mean, our objector persists, to describe abstract objects as 'somehow or other' responsible for our intuitions? Isn't this just more mystery-mongering? A fair question; it deserves an answer.

A mysterious process?

Mathematical intuition is mysterious because we know nothing at all about the perception of sets that Gödel, Hardy and other Platonists mention. Terms such as 'grasping', 'apprehension', and 'a kind of perception' are regularly used. But, it has been objected, this is quite unlike the situation regarding the five ordinary physical senses which we know something about. Dummett protests that the mathematical 'perception' which is alleged to exist has about it the 'ring of philosophical superstition' (1967, 202), and Benacerraf remarks that the nature of abstract objects 'places them beyond the reach of the better understood means of human cognition (e.g., sense perception and the like)' (1973, 409).

But how much more do we know about physical perception than mathematical intuition? In the case of ordinary visual perception of, say, a teacup, we believe that photons come from the physical teacup in front of us, enter our eye, interact with the retinal receptors and a chain of neural connections through the visual pathway to the visual cortex. After that we know virtually nothing about how beliefs are formed. The connection between mind and brain is one of the great problems of philosophy. Of course, there are some sketchy conjectures, but it would be completely misleading to suggest that this is in any way 'understood'. Part of the process of cognition is well understood; but there remain elements which are just as mysterious as anything the Platonist has to offer. Let's face it: we simply do not know how the chain of physical events culminates in the belief that the teacup is full. Of course, we should not glory in this state of ignorance. I suggest only that mathematical intuition is no more mysterious than the final link in physical perception. We understand neither; perhaps some day we will understand both.

Moreover, the correctness of our beliefs about the first part of the ordinary cognitive process (the part involving photons and signals to the visual cortex) is not in any way important to the issue at hand. Our present views about perception are highly theoretical and of only recent vintage. Those views might well be wrong, and what our ancestors did believe not so long ago about such processes is definitely wrong. For example, the onion skin theory of perception once prevailed. In this view, an infinitely thin surface (like a layer of an onion) came off an object and entered the perceiver's eye. Our present account of perception might be as wrong as the onion theory. Nevertheless, whether we are right or wrong in our present account of perception, we are (and our ancestors were) rightly convinced that there are material objects such as trees and tables which exist independently of us and that they somehow or other are responsible for our knowledge of them.

Whether we do or do not know the details of the interaction which takes place 'somehow or other' between abstract objects and ourselves is of secondary importance. Platonism is not to be disarmed because it has—as yet—no story to tell about the cognitive process of intuition.

The causal theory of knowledge

The objection to Platonism stemming from the causal theory is a very forceful line of argument which has been endorsed by many people starting with Benacerraf (1973) and including Lear (1977), Field (1980), Kitcher (1983), and Papineau (1987). There have been attempted rebuttals by several, including Steiner (1975), who only accepts a much weakened version of the causal theory which does no harm to Platonism; Maddy (1980), who avoids the threat by allowing only the perception of physical sets; Lewis (1986) who dismisses it as dubious epistemology, much less secure than the mathematics it challenges; and Wright (1983) and Hale (1987) who both thoroughly discuss the issue, and considerably lessen its plausibility, but do not forcefully come down against it. In spite of these responses, it seems fair to say that the causal theory of knowledge has largely won the day, and its use as a club to bash Platonists is widely accepted. It is a major ingredient in an increasingly prevalent naturalistic view of the world.

The popularity of the causal objection has been recent. One doesn't see it at all in the first part of the twentieth century. Nevertheless, this is certainly not its first appearance in history. One version of the objection is quite old, but may be of some interest.[8] Sextus Empiricus argued that the Stoics' notion of a proposition is unintelligible. Since the Stoics were thoroughgoing materialists about the mind, they were quite vulnerable to Sextus's objection.

> Let it be supposed and gratuitously conceded, for the sake of advancing our inquiry, that 'expressions' (i.e., what is said) are 'in existence', although the battle regarding them remains unending. If, then, they exist, the Stoics will declare that they are either corporeal or incorporeal. Now they will not say that they are corporeal; and if they are incorporeal, either—according to them—they effect something, or they effect nothing. Now they will not claim that they effect anything; for according to them, the incorporeal is not of a nature either to effect anything or to be affected. And since they effect nothing, they will not even indicate and make evident the thing of which they are signs; for to indicate anything and to make it evident is to effect something. But it is absurd that the sign should neither indicate nor make evident anything; therefore the sign is not an intelligible thing, nor yet a proposition.
>
> (*Against the Logicians* II, 262–4)

As we can see from this ancient example, there is actually a long tradition of using causal considerations to discredit abstract entities. Fortunately for the Platonist, the causal theory is false.

In order to show this, the precise details of the causal theory of knowing are not important. The spirit of any account worthy of the name will include the idea that to know about something one *must* have some sort of causal connection with the thing known. Among various discussions, the *locus classicus* is perhaps Alvin Goldman's essay, 'A Causal Theory of Knowing' (1967). There are numerous accounts which develop the theory or modify and apply it.[9] Undoubted is its appeal;

everyday examples of knowledge fit the bill nicely. For example, I know (at the time of writing) that it is snowing in Moscow. How do I know this? Photons are entering the eyes of someone there who telephones a newspaper office in Toronto causing a typesetter to do certain things resulting in the printed page in front of me. Photons from this page enter my eye making me believe that it is snowing in Moscow. There is a causal chain from the snow in Moscow to me, and if some link were missing then I probably wouldn't have had the belief I did.

Unless I'm causally linked somehow or other to the snow in Moscow I certainly don't know about it. Analogously, unless I'm causally linked somehow or other to abstract objects then I certainly won't know about them. It is, of course, a common presumption that none of us is causally linked to anything outside of space and time.

Among those who would like to rid themselves of the albatross of the causal theory is David Lewis. His modal realism takes quite literally the existence of possible worlds and treats them as every bit as real as ours. These worlds in no way causally connect with ours; so champions of the causal theory of knowledge then ask, How can we know anything about them? Lewis, of course, claims to know quite a bit, so a reply is called for. In defending himself he takes mathematics as a precedent. It should be understood in a literal and straightforward way, according to him.

> To serve epistemology by giving mathematics some devious semantics would be to *reform* mathematics . . . our knowledge of mathematics is ever so much more secure than our knowledge of the epistemology that seeks to cast doubt on mathematics. Causal accounts of knowledge are all very well in their place, but if they are put forward as *general* theories, then mathematics refutes them.
>
> (Lewis 1986, 109)

Those who insist on the causal theory won't be persuaded with moves which seem question begging. Without further argument they will not let mathematics be a precedent, and will instead insist on accounts of both mathematics and possible worlds that are compatible with the causal theory of knowing. To simply disavow the causal theory, as Russell once remarked in a different context, has all the advantages of theft over honest toil. Lewis, in dismissing the causal theory, shows the right instincts, but we can do a bit better; we can produce an argument for its rejection. Fortunately, the toil involved is not too great.

My case against the causal theory is drawn from recent results in the foundations of physics. For those who are familiar with the background, the argument is simply this: Given an EPR-type setup, if we know the outcome of a spin measurement on one side, then we can also know the outcome of a measurement on the other side. There is, however, no causal connection (either direct or linked through hidden variables) between these events. (This, at least on one interpretation, is the principal upshot of the so-called Bell results.) Consequently, we have knowledge without a causal connection, and so the causal theory of knowledge is false. For those unfamiliar with the background, I'll try to spell the whole argument out briefly.

First, some of the features of any causal theory. I said above that I wasn't concerned with details. That's because full details would spell out how a causal connection of the appropriate sort is *sufficient* for knowledge. All I'm interested in is defeating the common claim of any causal theory that some sort of causal connection is *necessary* for knowledge. I'll give only detail enough for that.

The first pattern of causal connection is the direct one which, schematically, we'll put this way (where 'sKp' means subject s knows that proposition p is true):

$$p \longrightarrow sKp$$

Examples abound. The fact that the cup is on the table before me causes me through ordinary sense perception to believe that the cup is on the table. (I'll talk indifferently about facts and events causing their effects. There may be important differences, but they are not at issue here.)

What about knowledge of the future? How can I know it will rain tomorrow, if we rule out backwards causation? The causal theorist has a ready answer. We appeal to the principle of the common cause and thereby establish a causal link to any future event. I am directly causally connected with present cloud conditions; these conditions cause rain tomorrow as well as my knowledge of that future event. Schematically:

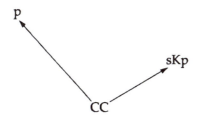

If the causal theory is generally recognized to have a problem, it is with generalizations. We seem to know that *all* ravens are black; but how do we manage this, given that we certainly aren't causally connected to each and every raven? Goldman thinks we know the general fact because we are causally connected to some instances. Schematically:

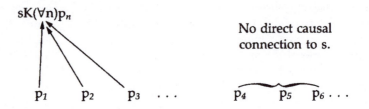

But, of course, this is extremely unsatisfactory for anyone who thinks we do have such general knowledge. Perhaps Popperians would be happy with this as an account of knowledge of generalizations—with the consequence they would gladly embrace that since we aren't in causal contact with all ravens we simply don't have the general knowledge that they are all black. Such generalizations should instead be seen as conjectures, on this view, not knowledge. However, those of us who are not sceptics will insist on something different.

> A better account of inference emerges if we replace 'cause' with 'because'. On the revised account, we infer not just statements of the form *X causes Y* but, more generally, statements of the form *Y because X* or *X explains Y*. Inductive inference is conceived as inference to the best of competing explanatory statements. Inference to a causal explanation is a special case . . . . Furthermore, the switch from 'cause' to 'because' avoids Goldman's *ad hoc* treatment of knowledge of generalizations. Although there is no causal relation between a generalization and those observed instances which provide us with evidence for the generalization, there is an obvious explanatory relationship. That all emeralds are green does not cause a particular emerald to be green; but it can explain why that emerald is green. And, other things being equal, we can infer a generalization only if it provides the most plausible way to explain our evidence.
>
> (Harman 1973, 130f.)

Harman calls this a 'modification' (1973, 141) of Goldman's causal theory of knowledge. If so, then the causal theory is no longer a threat to those doctrines I mentioned above, since it no longer demands a physicalistic cause. Platonic entities explain our mathematical knowledge; possible worlds explain our modal talk; and, as I shall argue below, relations among abstract universals explain observed regularities in the physical world. Of course, we shall continue to debate whether these are the *best* explanations, but no longer can such entities be ruled out in principle on the grounds that they cannot be causally connected to potential knowers.

On the other hand, there may be a modification of Goldman's account, different from Harman's, which is more in the spirit of the original. This modification simply holds that we are in causal contact with *all* instances of a generalization. Consider our knowledge that all ravens are black. True, we are directly connected to only a small number of ravens. But each of these is causally connected back to its distant ancestor, the 'first' raven, which in turn is causally connected forward to all other ravens. This causal connection could then be the ground of our knowledge of the universal proposition that all ravens are black.

Similarly, there might be something in the common past of each emerald which makes them all green. Thus our knowledge of the generalization is really the same as our knowledge of the rain tomorrow which is based on our being directly causally connected to clouds today.

At any rate, making sense of generalizations is the great stumbling block of the causal theory, but I don't want to make much of this problem here. The example, which I'll turn to now, will instead focus on the other two features of the causal theory.

Einstein, Podolsky, and Rosen (1935) set out an ingenious argument to undermine the then reigning interpretation of quantum mechanics, the Copenhagen interpretation, which holds that quantum mechanical systems have their properties *only* when measured. EPR attacked the idea that measurements *create* instead of *discover*. Since this has already been explained, I'll only briefly review it here. We start with a coupled system which can be separated, say an energetic particle which decays into a pair of photons moving in opposite directions. The initial system has spin 0 and spin is conserved. If we make a spin measurement on one photon along, say, the z axis, and get what is commonly called spin up, then a measurement on the other photon along the same axis will result in spin down. We further suppose that the two measurements are done far from one another, indeed, outside each other's light cones. This last requirement is sometimes called the *locality* assumption. According to special relativity, the upper bound on the velocity of any signal prevents the measurement of one photon having an effect on the other.

We know the outcome of the distant measurement, but how is this possible? How can a measurement on one side, EPR asks, *create* the spin components on the other, since there is no causal influence of one on the other? It must be that the spin components were already present at the origin of the system and that the measurements *discovered* what they were; observation does not create the photon's properties.

In terms of the causal theory of knowledge, we can see what is going on in the EPR argument. By making a measurement on one side we come to know the result by direct causal interaction. We also immediately know the other, but not through any direct causal connection. We come to know the distant outcome through a chain that goes to a common cause in the past of both measurements, something which caused the results in both cases. It is through this causal chain that we know of the remote result. It is analogous to our knowing today of the rain tomorrow.

EPR is an argument which concludes that the standard interpretation of quantum mechanics is incomplete; it could be completed by adding so-called hidden variables. The common cause at the origin which is causally responsible for the correlated outcomes would be such a hidden variable.

Alas, for all its brilliance, EPR does not work. J.S. Bell generalized the EPR argument, and subsequent experiments based on his results have shown that the EPR conclusion is hopeless. (The details of Bell's theorem are in chapter eight.) On the assumption that such hidden variables (i.e., the common cause) are present and that locality also holds, Bell was able to obtain an inequality that now bears his name. The remarkable thing is that his result has straightforward empirical consequences; it can be put to the test, something which we could never do with EPR. The outcome is not a happy one for so-called local realism. As long as we hold to special relativity—and we should probably do that—then we must disavow hidden variables. In any EPR-type situation *there is no common cause* of the correlated measurement outcomes.

The moral for the causal theory of knowledge is simple: it's wrong. We have knowledge of an event; but we are not causally connected in any physicalistic way with that event, either directly (because of locality) or through a common cause in the past (since there are no hidden variables).

It is important to see that the situation here is quite different from, say, my knowledge that some raven outside my light cone is black. There are similarities and differences, but the latter are crucial. The similarities are these: I am directly causally connected to some raven which is black and to some photon which has, say, spin up; I know the distant raven is black and the distant photon has spin down; I am not directly causally connected to the distant raven, nor to the distant photon. The crucial difference is this: I am indirectly causally connected to the distant raven via past ravens from whom they inherit their black colour. But I am *not* indirectly causally connected to the distant photon via anything at all that could cause the distant photon to be spin down. In accounting for this type of case (i.e., my knowledge of the spin state of the distant photon), the causal theory of knowledge is simply hopeless.

Nor is the situation to be trivialized by saying we know the distant spin measurement result because we know the theory. In a sense this is, of course, true. But the theory is not part of any causal chain from knower to thing known. Even if we did consider the theory as some sort of physical entity in its own right, still (my token of) the quantum mechanical theory is not (physically) connected to the spin components of the distant electron the way (my token of) the 'all ravens are black' theory could be said to be physically connected to each and every raven.

But just as no experiment in physics is really crucial, so no argument in philosophy is really conclusive. There is a respectable number of physicists, including David Bohm and Bell himself, who would rather give up special relativity than abandon any sort of causal connection between the distant outcomes. If this is done, a causal theorist could say that there is indeed a direct causal connection between the knower and the distant measurement result. The cost to physics, of course, is enormous, but the epistemic doctrine would be rescued. I am opting with the majority who refuse to pay the price. (The actual situation is more subtle than indicated here. The foregoing will be vindicated, however, by considerations in the final chapter.)

Ironically, the attitude toward the Bell results I'm adopting here is the same as that of the arch-anti-realist, Bas van Fraassen (1980, 1982). The realist (scientific realist, that is, not Platonist) argument he tries to undermine goes like this: Significant correlations must be explained by appeal to a common cause. Sometimes nothing observable can be found, so appeal must be made to something hidden (atoms, germs, genes, etc.) to explain the phenomena. Thus, the imperative to explain is sometimes an imperative to posit theoretical entities. In response to this argument, van Fraassen cites the Bell results as showing that the imperative cannot always be obeyed; there are significant correlations (i.e., spin correlations), but there cannot be a common cause which explains them. These correlations, according to van Fraassen, are just a brute fact about the world. To all of this I would agree, adding only that from the point of view of *physical causation* spin correlations are just a brute fact, but there is much more involved than this, as we shall see in the final chapter.

In sum, I've briefly characterized mathematical Platonism and tried to defend it from some of the charges which are commonly made against it, including the objection stemming from the causal theory of knowledge, which has in the past proven to be so persuasive. The debate between realists and nominalists has been long and glorious; there is little danger it will be settled now or in the near future. But we can continue to make a little progress here and there—by showing that our opponent's objections are misguided, if nothing more definite than that.

The real point of this chapter, however, is to set the stage for the next, where a Platonic account of thought experiments will be given. Acceptance of Platonism in mathematics will make the following account of thought experiments more enticing—or at least seem not so ridiculous. Many of the objections raised against mathematical Platonism have an analogue that will inevitably be raised against my view of thought experiments. This chapter nips them in the bud. Of course, I'd like to offer here a detailed account of the nature of mathematical intuition, since I could then carry it over to the epistemology of thought experiments, but in lieu of that I'm happy to be able to dispense (I hope) with some of the major objections to Platonism and to conclude that those of us who fancy we know something about abstract objects have nothing to fear from the nay-sayers of nominalism.

Even though we are finished with the main point of this chapter, why stop now? There is another topic that naturally and obviously arises— thought experiments in mathematics. Making the case for Platonism is one thing; making it for thought experiments inside mathematics is quite another. Visual reasoning in mathematics bears an obvious similarity to thought experiments in the natural sciences. We take this opportunity to explore some of those similarities and end with an example that illustrates their potential. Those with no interest in the nature of mathematics—a common but pitiable condition—may wish to skip the balance of this chapter.

Mathematical thought experiments

We can imagine someone saying: 'All this talk about thought experiments is dangerous rubbish. Scientists should stick to actual observation and established methods of deductive and inductive inference.' Almost no one is prepared to go so far as that when confronted with the great thought

experiments of physics, since they have obviously played a hugely important role. Yet a similar opinion is commonplace in mathematics. It is
part of the orthodox view that mathematics requires rigorous proofs,
and anything less is mere speculation. In particular, visual reasoning in
mathematics, involving the use of pictures, diagrams, or thought experiments, is usually considered illegitimate. This is the prevailing view. It
is usually conceded that pictures can provide useful insight and may
even suggest a proof, but they cannot serve as a substitute for a genuine
proof.

We should begin with examples to see how persuasive some of them
can be. The things to look for are these: Is the diagram convincing? Is it
a special case (i.e., for some particular *n*)? But does it establish complete
generality (i.e., for every *n*)? Would a standard verbal/symbolic proof of
the theorem, say, by mathematical induction, be more convincing than
the picture proof? Keep these questions in mind when considering the
following examples.

Theorem: $1 + 2 + 3 + \ldots + n = n^2/2 + n/2$

Proof:

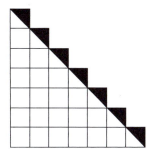

Figure 21

It may take a moment to see how the proof works. Start counting
from the top: $1 + 2 + 3 + \ldots + 7$. How many little squares are there? If we
squared the whole thing, then cut it with a diagonal, we would have the
number, except for the little squares along the diagonal, which got cut in
half. So add them back on. Now we see that the answer must be half of
7^2 plus half of 7. But this is a special case, where $n = 7$. Nevertheless, once
we understand the picture proof we see that it holds for any number, all
infinitely many.

This picture proof should be contrasted with a traditional proof by
mathematical induction which would run as follows:

Proof: We must show first, that the formula of the theorem holds for 1 (the basis step), and second, that if it holds for n, then it also holds for $n+1$ (the inductive step).

Basis step: $1 = 1^2/2 + 1/2$

Inductive step: Suppose $1 + 2 + 3 + ... + n = n^2/2 + n/2$ at least as far as n. Now we add the next term in the series, $n+1$, to each side, then calculate:

$$1 + 2 + 3 + ... + n + (n+1) = \frac{n^2}{2} + \frac{n}{2} + (n+1)$$

$$= \frac{n^2}{2} + \frac{n}{2} + \frac{2n}{2} + \frac{2}{2}$$

$$= \frac{n^2 + 2n + 1}{2} + \frac{n+1}{2}$$

$$= \frac{(n+1)^2}{2} + \frac{n+1}{2}$$

This last term has exactly the form we want. And so the theorem is proven.

Here is a second example, this time a theorem involving an infinite series.

Theorem: $\displaystyle\sum_{n=1}^{\infty} \frac{1}{4^n} = \frac{1}{4} + \frac{1}{16} + \frac{1}{64} + \frac{1}{256} + ... = \frac{1}{3}$

Proof:

Figure 22

The standard way to prove this theorem is with so-called ε-δ techniques (found in any advanced calculus textbook). But what about the diagram? One can see that the shaded areas correspond to the terms of the series. Can you also see where the 1/3 comes from? Is this picture proof convincing? Is it just as reliable as the standard proof?

With these examples under our belts, maybe we can turn our attention to an outstanding problem in mathematics, the continuum hypothesis, and try to solve it with a thought experiment.

What is the continuum hypothesis?

The continuum hypothesis (CH) was number one on David Hilbert's famous list of problems, most of which have now been settled. There are three ways of resolving a problem such as CH: prove it true, prove it false, prove it undecidable. CH, unfortunately, is the last of these. Before trying to explain any of this, let's quickly review some of the set theory basics.

The natural numbers, also known as the counting numbers, are in the set $\mathbf{N} = \{0, 1, 2, 3,...\}$. The size of this set, its *cardinality*, is infinite. Symbolically, $|\mathbf{N}| = \aleph_0$. What about other infinite sets, such as the set of even numbers, $\mathbf{E}$? How big is it? Two sets have the same cardinality if and only if there is a one-one, onto function between them. Such a map exists between $\mathbf{E}$ and $\mathbf{N}$. For instance: $1 \leftrightarrow 2, 2 \leftrightarrow 4, 3 \leftrightarrow 6, 4 \leftrightarrow 8,..., n \leftrightarrow 2n,...$ Thus, $|\mathbf{E}| = |\mathbf{N}|$. The set of rational numbers, the fractions, $\mathbf{Q}$, turns out to be the same size as well. So, we have $|\mathbf{E}| = |\mathbf{N}| = |\mathbf{Q}| = \aleph_0$. We might be tempted to think that all infinite sets are the same size, but this famously turned out to be not so. The set of real numbers, $\mathbf{R}$, also known as *the continuum*, the set of points on the line, is larger. This was proven by Cantor and is surely one of the greatest mathematical results of all time.

To prove that $|\mathbf{R}| > \aleph_0$ we need to show two things. First, we need to show that $|\mathbf{R}|$ is at least as big as $|\mathbf{N}|$. This is easy, since $\mathbf{N}$ is a proper subset of $\mathbf{R}$. So there must be at least as many members of $\mathbf{R}$ as of $\mathbf{N}$. The second thing to show is that there is no one–one mapping between $\mathbf{N}$ and $\mathbf{R}$. That would show that they can't be the same size. Putting these two facts together gives us Cantor's spectacular result that the real numbers are more numerous than the natural numbers.

Cantor's proof that there is no one–one correspondence is the appropriately named *diagonal argument*. We begin by assuming that there is a one–one, onto map between $\mathbf{N}$ and $\mathbf{R}$. In fact, we can even focus on just the points in the interval $[0, 1]$. Perhaps the one–one correspondence looks something like this:

$$0 \leftrightarrow .88491625...$$
$$1 \leftrightarrow .12548179...$$
$$2 \leftrightarrow .39271254...$$
$$3 \leftrightarrow .56469848...$$
$$...$$

Now let's construct a number r according to the following rule: pick the first number in the first decimal place and change it, say, by lowering the digit by 1. Now take the second number in the second decimal place and change it in the same way. Then the third number in the third place, and so on. Thus, we have $r = .7115...$

Since r is a real number in the set [0, 1], it should be on the list, because the list was assumed to contain *all* real numbers between zero and one. However, r cannot be on the list. It cannot be the first number on the list, since it differs in at least one decimal place, namely, the first. Similarly, it cannot be the second number on the list, since it differs in at least the second decimal place. In general, it cannot be the nth number on the list, since it differs in the nth decimal place. Thus, r is not on the list. No matter what mapping we choose between **N** and [0, 1], there will always be some number definable in the same diagonal way that r was defined, so there will always be something left off the list. There are more real numbers, or equivalently, more points on a line, than there are counting numbers, which is why sets the size of **R** or greater are called uncountable. Thus, there can't be a one–one, onto function between **R** and **N**, so |**R**| must be larger than |**N**|.

Power sets are bigger, but how much bigger? It's useful to have Cantor's theorem in mind, but we can revert to the special case of the real numbers when considering the question. **R** is an infinite set that is larger than **N**. But how big is it? Since each real number is an infinite decimal expansion, the set of real numbers is an infinite set of infinite numbers. This means that its cardinality is $2^{\aleph_0}$. In general, the cardinality of the power set of S is $2^{|S|}$. Cantor's theorem establishes a hierarchy of sets with infinite cardinalities: $\aleph_0 < 2^{\aleph_0}$. The interesting question he faced concerns the place of **R**, the continuum, in the hierarchy: $0 < 1 < 2 < ... < \aleph_0 < \aleph_1 < \aleph_2 < \aleph_3 ...$ Does $2^{\aleph_0} = \aleph_1$? Or does it equal $\aleph_2$? Or perhaps $\aleph_3$? Cantor's continuum hypothesis is the claim that |**R**| $= \aleph_1$, or equivalently, that $2^{\aleph_0} = \aleph_1$. The so-called generalized continuum hypothesis is the claim that $2^{\aleph_n} = \aleph_{n+1}$. If CH is false, then |**R**| might equal $\aleph_2$, or $\aleph_{374}$, or perhaps it might be larger than $\aleph_n$, for any finite n. Then again, it could be wrong because the whole of transfinite set theory is utter rubbish. All possibilities should be kept in mind.

Though the continuum hypothesis is usually expressed in terms of transfinite cardinal numbers, these concepts are not essential to the problem. It actually arises in a very simple way in standard analysis. CH is equivalent to the claim that every set of real numbers is equivalent (i.e., there is a one–one, onto correspondence) to a countable set of natural numbers or to the set of all real numbers.

The early twentieth century saw lots of failed attempts to prove or to refute CH. The first significant advance came in 1938 when Gödel proved that CH is *consistent* with the rest of set theory. He did this by providing a model based on the so-called constructable sets, in which all the axioms of ZFC are true and CH is true as well. This means, of course, that CH cannot be refuted in the normal way, that is, by proving ~CH via a derivation from the axioms of ZFC.

Full independence was established in 1963 by Paul Cohen. He introduced a powerful new technique called forcing that allowed him to construct a model of set theory in which ZFC is true but CH is not. The Gödel and Cohen results together establish undecidability. In other words, CH is *independent* of ZFC; it cannot be proven and it cannot be refuted—at least, not in the usual way. This is how things stand today. It leads to a pair of big questions.

First, does CH have a truth-value? Many people identify truth in mathematics with proof. We can't do that here, since no proof of CH (or of ~CH) is possible. It will come as no surprise that Platonists think that every mathematical proposition, including CH, is true or is false, whether we can prove it or not. This is the view I adopt here.

Second, given that a normal proof is not possible, how else might we decide the truth-value of CH and is that other way legitimate? It turns out that there may indeed be another way—a thought experiment, due to Christopher Freiling (1986). I am inclined to think his thought experiment works, but, I admit, it is not a sure thing.

Freiling's refutation of CH

Christopher Freiling (1986) produced a remarkable result that has gone largely unnoticed by philosophers. He refuted the continuum hypothesis. *Really* refuted the continuum hypothesis? That's hard to say. The standards for success in such a venture are not normal mathematical standards, since CH is independent of the rest of set theory. So any 'proof' or 'refutation' will be based on considerations outside current mathematics. Because of this, Freiling calls his argument 'philosophical'.

Some background. We shall take ZFC for granted. C, as I mentioned earlier, refers to the axiom of choice. An important consequence of ZFC is the so-called *well-ordering principle*. It says that any set can be well-ordered, that is, can be ordered in such a way that every subset has a first element. The usual ordering, $<$, on the natural numbers is also a well-ordering. Pick any subset, say $\{14, 6, 82\}$, it has a first element, namely 6. But the usual ordering on the real numbers is not a well-ordering. The subset $(0, 1) = \{x: 0 < x < 1\}$, for instance, does not have a first element. Nevertheless, ZFC guarantees that the real numbers can be well-ordered by some relation, $\prec$, even though no one has yet found such a well-ordering. Now we can turn our attention back to CH.

Imagine throwing darts at the real line, specifically at the interval $[0,1]$. Two darts are thrown and they are independent of one another. The point is to select two random numbers. (Obviously, this is a thought experiments, since we could not pick out real numbers, which are mere points, with physical darts.) We assume ZFC and further assume that CH is true. Thus, the points on the line can be well-ordered so that for each $q \in [0, 1]$, the set $\{p \in [0, 1]: p \prec q\}$ is countable. (Note that $\prec$ is the *well-ordering relation*, not the usual *less than*, $<$.) The well-ordering is guaranteed by ZFC; the fact that the set is countable stems from the nature of any ordering of any set that has cardinality $\aleph_0$. To get a feel for this, imagine the set of natural numbers. It is infinite, but if you pick a number in the ordering, there will be only finitely many numbers earlier—and infinitely many numbers later. Similarly, pick a number in a well-ordered set that is $\aleph_1$ in size and you will in effect define a set of earlier members that is at most $\aleph_0$ in size and possibly even finite. In any case, it will be a countable set.

We shall call the set of elements that are earlier than the point q in the well-ordering S_q. Suppose the first throw hits point q and the second hits p. Either $p \prec q$, or vice versa; we'll assume the first. Thus, $p \in S_q$. Note that S_q is a *countable* subset of points on the line. Since the two throws were independent, we can say the throw landing on q defines the set S_q 'before' or 'independently from' the throw that picks out p. The measure of any countable set is 0. So the probability of landing on a point in S_q is 0. (It is not important to understand the notion of *measure*, only the consequence for probability theory, to which it is linked.)

The same argument can be made for the set S_p, that is, there is zero chance that the other dart will land on a point in that set. Nevertheless, one of the two darts will land in the countable set defined by the other dart. While logically possible, this sort of thing is almost never the case. Yet it will happen every time there is a pair of darts thrown at the real line. Consequently, we should abandon the initial assumption, CH, since it lead to this absurdity. Thus, CH is refuted and so the number of points on the line is greater than $\aleph_1$.

Notice that if the cardinality of the continuum is $\aleph_2$ or greater, then there is no problem (at least as set out here), since the set of points S_q

earlier in the well-ordering need not be countable; it would not automatically lead to a zero probability of hitting a point in it.

Freiling actually goes on to show that there are infinitely many cardinal numbers between $\aleph_0$ and the continuum. Three darts will refute $\aleph_2$, four darts will refute $\aleph_3$, and so on. It would seem that the continuum cannot be $\aleph_n$ for any finite number n. Speaking as a Platonist, this strikes me as a very plausible result.

There is one aspect of this example on which I should elaborate. The darts give us a pair of real numbers picked at random. These are 'real random variables', says Mumford (2000), whose version of the thought experiment I have followed. The concept of random variable at work here is not the mathematical concept found in measure theory (a defined concept inside set theory that will not yield ~CH). Moreover, the two real numbers are picked independently; either could be considered as chosen 'first'. This means that the example cannot be dismissed in the way we might dismiss someone who said of a licence number, say, 915372, on a passing car: 'Wow, there was only a one chance in a million of that!' We're only impressed if the number is fixed in advance. The independence and randomness of the darts guarantees the symmetry of the throws, so either could be considered the first throw that fixes the set of real numbers that are earlier in the well-ordering.

An example such as this must be controversial. Only a minority of mathematicians have accepted it. Freiling himself (personal correspondence) has come to the view that it is not CH that has been refuted, but rather the well-ordering principle. He may be right, though I'm inclined to see CH as the thing that has to go. But for our purposes here, it does not matter. Well-ordering, as I noted earlier, follows from the axiom of choice, which is also independent from the rest of set theory. So its status is similar to that of CH, namely, it is neither provable nor refutable (in the normal way). The moral of this case is that a major mathematical principle (CH or well-ordering), which is independent from the rest of set theory, has been settled by a thought experiment. Ultimately, the dart thought experiment may be considered illegitimate, but for now it is certainly plausible and the onus is on critics to tell us why it doesn't work. The use of thought experiments, so profoundly and obviously effective in physics, may prove similarly valuable in mathematics.

5 Seeing the laws of nature

Happy are they who see the causes of things—Virgil

The existence of a special class of thought experiments—*platonic*—was asserted earlier when the taxonomy was created. This chapter is an attempt to vindicate that claim. I shall go about this by arguing, first, that these thought experiments are indeed *a priori*; and second, that laws of nature are relations between independently existing abstract entities. The existence of such entities gives the thought experimenter something to perceive. It also makes obvious which sense of *a priori* is at work; it is the same as that involved in mathematical Platonism. Neither linguistic conventions nor Kantian forms of perception—both candidate accounts of the *a priori*—are involved. Some laws of nature, on my view, can be seen in the same way as some mathematical objects can be seen.

Of course, no case for any theory can be made except in comparison with alternative accounts; so, scattered throughout this chapter as in earlier chapters are remarks on the rival views of Kuhn, Mach, and Norton. The relative merits of alternatives should always be kept in mind.

What are Platonic thought experiments?

Thought experiments in physics work in a variety of ways. As we saw in the second chapter, some do their job in a *reductio ad absurdum* manner by destroying their targets. Others are constructive; they establish new results. A very small number of thought experiments seem to do both—in destroying an earlier theory they also bring a new one into being. These are the Platonic thought experiments. We can characterize them this way:

A *Platonic thought experiment* is a single thought experiment which destroys an old or existing theory and simultaneously generates a new one; it is *a priori* in that it is not based on new empirical evidence, nor is it merely logically derived from old data; and it is an advance in that the resulting theory is better than the predecessor theory.

The best examples of this are Galileo's thought experiment concerning the rate of fall of bodies of different weights, the EPR thought experiment which destroyed the Copenhagen interpretation and established hidden variables, and Leibniz's argument for the conservation of *vis viva*. (Remember that these are all fallible, in spite of being impressively ingenious.)

Why *a priori*?

Let's quickly review the first of these examples. Galileo asks us to imagine a heavy ball (H) attached by a string to a light ball (L). What would happen if they were released together? Reasoning in the Aristotelian fashion leads to an absurdity. The lighter ball would slow down the heavy one, so the speed of the combined balls would be slower than the heavy ball alone (H + L < H). However, since the combined balls are heavier than the heavy ball alone, the combined object should fall faster than the heavy one (H < H + L). We have a straightforward contradiction; the old theory is destroyed. Moreover, a new theory is established; the question of which falls faster is obviously resolved by having all objects fall at the same speed.

Galileo's thought experiment is quite remarkable and we are justified in calling this a case of *a priori* knowledge. Here's why:

(1) *There have been no new empirical data.* I suppose this is almost true by definition; being a *thought* experiment rules out new empirical input. I think everyone will agree with this; certainly Kuhn (1964) and Norton (1991) do. It's not that there are no empirical data involved in the thought experiment. The emphasis here should be on *new* sensory input; it is this that is lacking in the thought experiment. What we are trying to explain is the *transition* from the old to the new theory and that is not readily explained in terms of empirical input unless there is new empirical input. (I will deal with the idea of 'old data seen in a new way' below when discussing Kuhn's view.)

(2) *Galileo's new theory is not logically deduced from old data. Nor is it any kind of logical truth.* A second way of making new discoveries— a way which does not trouble empiricists—is by deducing them from

old data. Norton holds such a view when he claims that a thought experiment is really an argument. As we saw in the second chapter, this view will certainly not do justice to all thought experiments.

But might it account for those I call Platonic? I think not. The premises of such an argument could include all the data that went into Aristotle's theory. From this Galileo derived a contradiction. (So far, so good; we have a straightforward argument which satisfies Norton's account.) But can we derive Galileo's theory that all bodies fall at the same rate from these same premises? Well, in one sense, yes, since we can derive anything from a contradiction; but this hardly seems fair.[1] What's more, whatever we can derive from these premises is immediately questionable since, on the basis of the contradiction, we now consider our belief in the premises rightly to be undermined.

Might Galileo's theory be true by logic alone? To see that the theory that all bodies fall at the same rate is not a logical truth, it suffices to note that bodies might fall with different speeds depending on their colours or on their chemical composition (as has recently been claimed by Fischbach *et al.* 1986). The likely reason Galileo's thought experiment works for light/heavy but not for colours is that the former are additive or extensive, while the latter are not (i.e., combining two red objects will not make a redder object).

These considerations help to undermine the argument view of thought experiments.

(3) *The transition from Aristotle's to Galileo's theory is not just a case of making the simplest overall adjustment to the old theory.* It may well be the case that the transition was the simplest, but that was not the reason for making it. (I doubt that simplicity or other aesthetic considerations ever play an important role in science, but for the sake of the argument, let's allow that they could.)[2] Suppose the degree of rational belief in Aristotle's theory of falling bodies is r, where $0 < r < 1$. After the thought experiment has been performed and the new theory adopted, the degree of rational belief in Galileo's theory is r', where $0 < r < r' < 1$. That is, I make the historical claim that the degree of rational belief in Galileo's theory was *higher* just after the thought experiment than it was in Aristotle's just before. (Note the times of appraisal here. Obviously the degree of rational belief in Aristotle's theory *after* the contradiction is found approaches zero.) Appeals to the notion of smallest belief revision won't even begin to explain this fact. We have not just a new theory—we have a better one.

As well as these there are other reasons which suggest the example yielded *a priori* knowledge[3] of nature, but possibly the most interesting and most speculative has to do with its possible connection to a realist account of laws of nature recently proposed by Armstrong, Dretske, and Tooley.

Laws of nature

Thought experiments often lead to laws, but what are laws of nature anyway? There are two main contenders. The long-established view follows Hume in holding that a law is nothing more than a regularity; laws are supervenient on physical happenings. The other view is an upstart—though it is as old as Plato. It holds that laws are relations among universals, that is, connections between independently existing abstract entities. In this and the next section I will argue against Hume and for Plato.

Hardly any philosophers follow David Hume into general scepticism, but few philosophical doctrines have prevailed to such an extent as Hume's view of causality and the laws of nature. 'All events seem entirely loose and separate', says Hume. 'One event follows another, but we never can observe any tie between them' (*Enquiry*, 74).'. . . [A]fter a repetition of similar instances, the mind is carried by habit, upon the appearance of one event, to expect its usual attendant, and to believe that it will exist' (*Enquiry*, 75). Causality and the laws of nature are each nothing more than regularities. To say that fire causes heat or that it is a law of nature that fire is hot, is to say nothing more than that fire is constantly conjoined with heat. Hume defined cause as 'an object, followed by another, and where all the objects similar to the first are followed by objects similar to the second' (*Enquiry*, 76).[4] We can't see a 'connection' between fire and heat such that if we knew of the one we could know that the other must also occur. All we know is that whenever in the past we have experienced one we have also experienced the other. Hence, the 'regularity' or 'constant conjunction' view of causality and laws of nature.

The appeal to empiricists is evident. All that exists are the regular events themselves; there are no mysterious connections between events—no metaphysics to cope with. The general form of a law is simply a universal statement. It is a law that As are Bs has the form $(x)(Ax . Bx)$. 'It is a law that ravens are black' comes to: all ravens are black; 'it is a law that quarks have fractional charge' comes to: all quarks have fractional charge.

But the elegance of Hume's account is somewhat mitigated by a multitude of examples which have the universal form, yet clearly are not laws of nature. First, consider vacuous truths:

All unicorns are red.
All unicorns are not red.

These would have to be counted as laws on the naive version of the regularity account. They are vacuous truths since there are no unicorns, but they are truths just the same, since any universal conditional with a false antecedent is automatically true. Thus, there would be all kinds of bizarre laws.

An apparently simple way to get rid of them is to disallow sentences with false antecedents from the class of laws. However, Newton's first law which says that a body which is acted upon by no force remains in a state of rest or constant rectilinear motion is vacuously true. In our universe where every body gravitationally interacts with every other body, none is free of feeling some force.

Let's turn to some non-vacuous generalizations which we'll suppose true:

All apples are nutritious.
All silver conducts electricity.
All the fruit in the basket are apples.
All the coins in Goodman's pocket on VE Day are silver.

What is the difference between genuine laws (the first two are likely candidates) and 'accidental' generalizations (the last two)? Almost all champions of the regularity view have admitted some sort of difference, but they have all wanted to place that difference *in us*, not in nature.

Those who adopt such a subjective view of laws of nature include: Nelson Goodman: 'we might say a law is a true sentence used for making predictions. . . . [R]ather than a sentence being used for prediction because it is a law, it is called a law because it is used for prediction . . .' (1947, 20); and A.J. Ayer: 'My suggestion is that the difference between our two types of generalization lies not so much on the side of the facts which make them true or false, as in the attitude of those who put them forward' (1956, 88).

If different people were to adopt different attitudes to the various generalizations it would mean they had different laws—and no one could be said to be wrong. This bizarre consequence is even admitted by R.B. Braithwaite, another champion of the subjective view, who says his 'thesis makes the notion of natural law an epistemological one and makes the "naturalness" of each natural law relative to the rational corpus of the thinker' (1953, 317).

As well as the relativity of laws on this subjective account, there is another unpalatable consequence. Before there were sentient beings who could adopt different attitudes to various generalizations, there were no laws of nature at all. Empiricists are aware of these consequences and are prepared to accept them—I'm not.

It seems logically possible that a world could be governed by laws and yet have no (repeatable) regularities. To contend with this and other difficulties in the original Humean account John Earman has proposed the 'empiricist loyalty test' which involves subscribing to the principle: if two possible worlds agree on the occurrent facts then they agree on the laws. He rightly claims that this principle 'captures the central empiricist intuition that laws are parasitic on occurrent facts' (Earman 1986, 85). A variation on the Humean theme which does justice to Earman's loyalty test has been proposed by John Stuart Mill, by Frank Ramsey, and (following Ramsey) by David Lewis. Laws, on this account, are propositions at the heart of any systematization of the facts of nature (regularities or not). Ramsey: 'causal laws [are] consequences of those propositions which we should take as axioms if we knew everything and organized it as simply as possible in a deductive system' (1931, 242). Ramsey only held the view for a short time, since 'it is impossible to know everything and organize it in a deductive system', But David Lewis (1973) pointed out that this is a poor reason—we can talk about the ideal systematization (one which best combines simplicity and strength) without knowing what it is.

This view overcomes the objections raised above that the laws are relative to belief systems or that they are non-existent before humans came on the scene, since ideal systematizations exist quite independently of us. Nevertheless, while the problems stemming from subjectivity are overcome, others remain.

Consider the following situation which is inspired by an example from Tooley (1977). Imagine that the Big Bang took place in two stages. In the first, a class of particles—let's call them 'zonks'—receded from the rest of the remaining matter in the universe at nearly the speed of light. And let us further suppose that this separation of types of matter is governed by quantum mechanical laws which are only statistical; that is, the separation isn't necessary. Thus, never in the entire history of the universe would zonks interact with protons, electrons, etc. It seems reasonable to say, however, that even though zonks and protons never interacted, if they had they would have done so in a law-like way. That is, there are laws governing the interaction of zonks and protons. However, any systematization of the regularities of this universe would include only vacuous truths about zonk-proton interactions. Thus, we would have:

In zonk-proton collisions plonks are emitted.
In zonk-proton collisions no plonks are emitted.

One and only one of these is a law of nature. Thus, the Mill-Ramsey-Lewis-Earman account can't do justice to this situation. There is more to a law of nature than a regularity or a set of occurrent facts, and that something more is not captured either by subjective attitudes or by ideal deductive systematizations—that something extra must be in reality itself.

Laws as relations among universals

A new account of laws has been proposed in the light of great dissatisfaction with any regularity view. It is the simultaneous, independent creation of David Armstrong, Fred Dretske, and Michael Tooley. Each claims that laws of nature are relations among universals, that is, among abstract entities which exist independently of physical objects, independently of us, and outside of space and time.[5] It is a species of Platonism.

The 'basic suggestion', according to Tooley, 'is that the fact that universals stand in certain relationships may logically necessitate some corresponding generalization about particulars, and that when this is the case, the generalization in question expresses a law' (1977, 672).

A law is not a regularity, it is rather a link between properties. When we have a law that Fs are Gs we have the existence of universals, F-ness and G-ness, and a relation of necessitation between them. (Armstrong symbolizes it: N(F,G).) A regularity between Fs and Gs is said to hold in virtue of the universals F and G. '[T]he phrase "in virtue of universals F and G" is supposed to indicate', Armstrong says, that 'what is involved is a real, irreducible, relation, a particular species of the necessitation relation, holding between the universals F and G . . .' (1983, 97).

The law entails the corresponding regularity, but is not entailed by it. Thus we have:

$$N(F,G) \rightarrow (\forall x)(Fx \supset Gx)$$

And yet:

$$(\forall x)(Fx \supset Gx) \not\rightarrow N(F,G)$$

The relation N of nomic necessity is understood to be a primitive notion. It is a theoretical entity posited for explanatory reasons. N is also understood to be contingent. At first sight this seems to be a contradiction. How can a relation of necessitation be contingent? The answer is simple. In this world Fs are required to be Gs, but in other worlds Fs may be required to be something else. The law N(F,G) is posited only for this world; in other possible worlds perhaps the law N(F,G') holds.

The new view has lots of precursors. One of the most interesting is C.S. Peirce.[6] Though in many respects a staunch empiricist, he felt driven to acknowledge the reality[7] of what he called 'thirdness', a category in his ontology of things, which include the laws of nature.

> With overwhelming uniformity, in our past experience, direct and indirect, stones left free to fall have fallen. Thereupon two hypotheses only are open to us. Either—
>
> 1. the uniformity with which those stones have fallen has been due to mere chance and affords no ground whatever, not the slightest, for any expectation that the next stone that shall be let go will fall; or
> 2. the uniformity with which stones have fallen has been due to some *active general principle*, in which case it would be a strange coincidence that it should cease to act at the moment my prediction was based upon it.
>
> Of course, every sane man will adopt the latter hypothesis. If he could doubt it in the case of the stone . . . a thousand other such inductive predictions are getting verified every day, and he will have to suppose every one of them to be merely fortuitous in order reasonably to escape the conclusion that *general principles are really operative in nature*. That is the doctrine of scholastic realism.
>
> (Peirce 1931–35, vol. V, 67)

Some of the advantages of a realist view of laws are immediately apparent. To start with, this account distinguishes—objectively— between genuine laws of nature and accidental generalizations. Second, laws are independent of us—they existed before we did and there is not a whiff of relativism about them. Third, even if there is no interaction between kinds of particle (zonks and protons) there are still laws governing those (possible) interactions. Finally, uninstantiated laws (e.g., Newton's first law) are not merely vacuously true, but are (candidates for) genuine laws.

There is lots more to be said about the problems of the regularity view of laws, lots more to be said about laws as relations among universals, and lots more to be said about the virtues of such a Platonic account. Armstrong (1983), Dretske (1977), and Tooley (1977) must be consulted for more. But before leaving this section I will add one final epistemological point. One might wonder whether all this mysterious metaphysics isn't a bit hard to test. Wouldn't an empiricist account make specific conjectured laws of nature easier to evaluate than laws which are encumbered with Platonic trappings?

> Another reason for starting with the constant conjunction view is that, according to it, scientific laws are logically weaker propositions than they would be on any alternative view of their nature. On any other view a scientific law, while including a generalization, states something more than the generalization. Thus, the assumption that a scientific law states nothing beyond a generalization is the most modest assumption that can be made. . . . It is difficult enough to justify our belief in scientific laws when they are regarded simply as generalizations; the task becomes more difficult if we are required to justify belief in propositions which are more than generalizations.
>
> (Braithwaite 1953, 11)

Braithwaite's point seems compelling initially, but on consideration collapses. If science were really interested in discovering generalizations it would be trying to determine whether all the coins in Goodman's pocket on VE Day were silver. On the contrary, science is after the laws. And even for modern followers of Hume, including Braithwaite, those laws will be regularities plus something else—though that extra something will not have to do with the world. To establish that Fs are Gs is a law, a modern Humean will have to establish two things:

(1) $(\forall x)(Fx \supset Gx)$
(2) That (1) plays the right sort of role in either *the ideal* or *our present* systematization of the facts of the world.

The second of these conditions is hopeless. First, how could we ever know that (1) is included in the final science? It's hard enough merely to establish rational belief for here and now. The second possibility, establishing that it plays the right role in our present scheme of things would seem easy—we merely have to ask ourselves if we consider, say, F=ma a law or not. Alas, it's not always so easy.

Consider a very complicated molecule and a well-verified conjecture about its energy levels. Thus, we know that the appropriate generalization is true, but is it a law? To know whether it plays the right sort of role in *our* present scheme is to derive the energy levels from the Schrödinger equation. For a suitably complex molecule it is hard enough even to write down the Schrödinger equation; an exact solution is humanly out of the question. Even a numerical solution using supercomputers could take more than the entire history of the universe to achieve. Confirmation is impossible—modest empiricism is not so modest after all.

Laws and thought experiments

Here is not the place to argue at length for the merits of a Platonic account of laws. That has already been done admirably by Armstrong, Dretske, and Tooley. I want only to adopt it and to point out the harmony between it and my *a priori* account of thought experiments. I want to suggest that the way some thought experiments work—the Platonic ones—is by allowing us to grasp the relevant universals. The epistemology of thought experiments is similar to the epistemology of mathematics. Just as we sometimes perceive abstract mathematical entities, so we sometimes perceive abstract universals.

Let me put this another way. Suppose Gödel is right: we can 'see' some mathematical objects (which are abstract entities). And suppose the Armstrong-Dretske-Tooley account of laws of nature is also right: laws are relations among universals (which are abstract entities). Wouldn't it be a surprise and indeed something of a mystery if we couldn't 'see' laws of nature, as well? Isn't the ability to grasp them just as we grasp mathematical objects exactly what we should expect?

The reason this issue of the nature of laws is so important to my account of thought experiments is simply this: I need something for thought experimenters to see. So far I've argued for the *a priori* nature of (some) thought experiments, but the term '*a priori*' has several different interpretations—only one is to my liking.

The linguistic interpretation has it that *a priori* truths are true merely in virtue of the meanings of the terms involved. Thus, 'Bachelors are unmarried males' is known to be true independently of experience, because the very meaning of 'bachelor' is 'unmarried male'. Sensory experience plays no role since it is a truth about language, not about the world. A second interpretation of *a priori* knowledge has it that it is innate, placed in the mind by God, or by an evolutionary process; or as Kant would have it, *a priori* knowledge has something to do with the structure of thought.

None of these accounts of the *a priori* should be believed, at least, not if they are intended to be exhaustive; nor should Plato's view (also innatist) be accepted in full. Plato held that our immortal souls once gazed upon the heavenly forms. Our *a priori* knowledge is the result of remembering what we forgot in the rough and tumble of birth.

The only part of this I wish to retain is (the non-innatist part) that universals (properties and relations) have an existence of their own and like mathematical objects can be grasped by the human mind. This is an objective view of *a priori* knowledge—it posits a non-sensory perception of independently existing objects.

If the empiricist account of laws can be undermined and the realist view of Armstrong, Dretske, and Tooley established, then it will contribute considerably toward vindicating the Platonic account of thought experiments.

Objections and replies

Many of the objections that spring to mind are analogous to standard objections to mathematical Platonism. I included the chapter on mathematical thinking in the hopes of nipping these objections in the bud and hence of easing the way into the present chapter—Platonism in mathematics is easier to swallow than *a priori* physics.

There is no need to go through each of these objections in detail. I will just briefly recap a couple for illustration; those who are interested can easily reformulate other objections and replies for themselves.

First objection: Even if the laws of nature exist in Platonic fashion, they are unknowable. To know x we must be in some sort of physical contact with x, but this cannot happen with abstract objects which are outside of space and time.

Reply: In an EPR-Bell set-up we have knowledge of the remote measurement result, yet there is no physical causal connection. Thus, physical causal connections are not necessary for knowledge.

Second objection: The analogy between physical perception and the intuition of abstract objects is weak since ordinary sense perception is well understood while the perception of laws of nature is a complete mystery.

Reply: The claim that we understand ordinary sense perception is simply fraudulent. At best we understand part—the *physical* process starting with photons emitted by an object and ending with neural activity in the visual cortex. From there to *belief* about the object seen is still a complete mystery. The perception of abstract laws of nature is certainly no more mysterious than that. Moreover, even if we didn't have any idea at all about how ordinary physical perception worked (remember, our ancestors certainly had the wrong idea), we still would be justified in believing in the existence of physical objects as the things we actually see. Similarly, we are justified in believing in the real existence of laws of nature as the objects of our intuitions.

As I mentioned above, I hope I have anticipated most of the major objections to a platonic account of thought experiments in the preceding chapter. However, there is one more issue I wish to consider. The Platonic account of thought experiments flies in the face of everything an empiricist holds dear.

> there is only one non-controversial source from which [thought-experimental] information can come: it is elicited from information we already have by an identifiable argument, although that argument might not be laid out in detail in the statement of the thought experiment. The alternative to this view is to suppose that thought experiments provide some new and even mysterious route to knowledge of the physical world. Thus Brown (1986, pp. 12–13) argues that thought experiments are a special window through which we can grasp the universals of an Armstrong-like account of physical laws. I can see no benefit in adopting a mysterious window view of thought experiments, when all the thought experiments I shall deal with (and have seen elsewhere) in modern physics can be readily reconstructed as arguments.
>
> (Norton 1991, 129)

The style of Norton's objection to taking thought experiments as a unique way of gaining knowledge is reminiscent of the objection of Benacerraf and others to Gödel's mathematical intuitions. Each claims the respective proposal is 'mysterious' and that there is an uncontroversial *reconstruction*—a formal derivation from axioms in the mathematical case and a formal derivation from empirical premises in the thought experiment case—yielding the same results as the mathematical intuition or the thought experiment but without the epistemological problems.

Perhaps the well-entrenched distinction between the so-called 'context of discovery' and the 'context of justification' lies implicitly behind Norton's belief that what really matters is the reconstructed argument. It is also implicit in Benacerraf's challenge to mathematical Platonism.

Discovery *vs* justification

Indeed, I suspect this distinction is in the back of the minds of all those who would reject a serious role for intuitions in mathematics or for thought experiments in physics. Critics are often willing to allow a psychological role for each, but when it comes to justification—real evidence, rational grounds for belief—then only the so-called reconstruction counts. It is the formal derivation (a Benacerraf might say) that is the *real proof* of the mathematical theorem; it is the empiricist reconstruction (a Norton might say) that is the *real evidence* of the physical theory. This is certainly Carl Hempel's view:

> [T]heir heuristic function is to aid in the discovery of regular connection. . . . But, of course, intuitive experiments-in-imagination are no substitute for the collection of empirical data by actual experimental or observational procedures.
>
> (1965, 165)

The discovery/justification distinction works well with examples such as Kekulé's famous dream of a snake biting its own tail which suggested to him that the benzene molecule is a ring. But it does not work well with the examples discussed here.

To put it picturesquely: first, imagine Galileo at a conference; he announces that all bodies fall with the same speed. Asked for empirical evidence, he admits to having none, but instead describes his thought experiment. Second, imagine the discoverer of the no-tiling theorem (discussed in the second chapter) announcing this result at the conference. Asked for a formal derivation, the discoverer of the theorem admits to not having one, but instead gives the 'short proof'. Now imagine Kekulé at this same conference. He claims the benzene molecule is a ring. Asked for empirical evidence, he says that he, too, has none; but—not to worry—he had a great dream about a snake.

Are these really on a par? What is perhaps going on is this: the 'short proof' of the tiling theorem is evidence of the existence of a formal derivation; the thought experiment is evidence of the possibility of an empiricist reconstruction. Kekulé's dream, on the other hand, is *not* evidence that empirical confirmation is just down the road. What I want to assert now is just this rather obvious fact: *Evidence is transitive.*[8] If P is evidence for Q, and Q is evidence for R, then P is evidence for R. Thus, if the 'short proof' is evidence for the existence of a formal derivation and the formal derivation is evidence for the theorem, then the 'short proof' is evidence for the theorem. Similarly, Galileo's thought experiment (unreconstructed) is evidence for his claim that all bodies fall with the same speed.

At the very least, abstract entities explain our beliefs as grounded by the initial (unreconstructed) evidence and our further belief that there is 'real evidence' to be had. Kekulé's dream has to be explained in some completely different way. Perhaps Freud had an answer.

Kuhn's paradigms

Platonic thought experiments are not just cases of seeing old empirical data in a new way; yet such a view is essentially Kuhn's.[9] In his extremely interesting and insightful essay on thought experiments (Kuhn 1964), he does not use the terminology of 'paradigms' and 'gestalt shifts' found in *The Structure of Scientific Revolutions,* but the ideas are the same. The thought experiment shows us a problem in the old framework; it induces a crisis, and this, says Kuhn, helps us to see the *old* data in a new way—re-conceptualized. A thought experiment generates a new paradigm. Kuhn is half sympathetic with a view he considers as traditional among philosophers, and describes as follows:

> Because it embodies no new information about the world, a thought experiment can teach nothing that was not known before. Or, rather, it can teach nothing about the world. Instead, it teaches the scientist about his mental apparatus. Its function is limited to the correction of previous conceptual mistakes.
>
> (Kuhn 1964, 252)

Kuhn no sooner outlines this view than he goes on to qualify it by remarking that 'from thought experiments most people learn about their concepts and the world together' (1964, 253). The sense in which we can learn about the world has to do with correcting previous conceptual 'mistakes' in some central concept. 'But', he stresses,

> we cannot, I think, find any intrinsic defect in the concept by itself. Its defects lay not in its logical consistency but in its failure to fit the full fine structure of the world to which it was expected to apply. That is why learning to recognize its defects was necessarily learning about the world as well as about the concept.
>
> (1964, 258)

Recall that for Kuhn there is no world which exists independently of any conceptualization we may have of it; the world is paradigm dependent. In one of his most dramatic claims he remarked, 'In a sense that I am unable to explicate further, the proponents of competing paradigms practice their trades in different worlds' (1970, 150). So the sense in which we learn about the world must be a highly qualified one.

Though extremely perceptive in many ways, Kuhn's views on thought experiments are ultimately not persuasive. There are a number of reasons for this. To start with, there are several thought experiments which have nothing to do with detecting problems in an old theory (e.g., Stevin's inclined plane or Newton's bucket). Yet, he makes the detection of problems central to all thought experiments.

In paradigm change, on Kuhn's view, there is no new paradigm that is *uniquely* and *determinately* the one that must be adopted. Yet Galileo's theory that all bodies fall at the same rate seems the unique belief one ought to adopt after Aristotle's theory in the light of the damage done to it by the thought experiment.

Moreover, even though Kuhn is generally right about the difficulties of comparing different paradigms, incommensurability problems do not seem to be present in the Galileo case. There has been no change of meaning in the terms 'light', 'heavy', and 'faster'. Galileo and his Aristotelian opponents are not talking past one another *during* the performing of the thought experiment. Indeed, it can only be performed because they do mean the same things by their common terms.

Ian Hacking points out another feature of thought experiments to which Kuhn's account may fail to do justice. This has to do with their somewhat eternal appeal. The famous square-doubling example in Plato's *Meno* continues to impress modern readers—even though we are 2,500 years away from it. This is true for a large number of other thought experiments as well. By tying them so tightly, as Kuhn does, to the fortunes of particular conceptual schemes—often long-departed ones—it becomes something of a mystery why they still have the power to impress us so deeply.

Tamar Zabo Gendler (1998) has proposed a sophisticated view that is similar in some regards to Kuhn's account. She is, by the way, also critical of Norton's argument view. Thought experiments, according to her (as with Kuhn),

> provide contexts within which sense can be made of previously incomprehensible conceptual distinctions. This happens when two features that are constantly conjoined in our representations of all actual cases are imaginatively separated in the thought-experimental scenario in a way that shows them to have been isolatable all along.
> (1998, 413)

This is surely right for a number of cases, but there are (as there were with Kuhn) many cases that do not fit this account (e.g., Steven's inclined plane and Newton's bucket). Gendler also claims that thought experiments gives us genuine *knowledge*.

> ... thought experiments rely on a certain *constructive participation* on the part of the reader, and the justificatory force of the thought

experiment actually comes from the fact that it calls upon the reader to perform what I will call an *experiment-in-thought*.

(1998, 413f.)

In other words, this is an actual experiment carried out by the reader who poses to herself the question 'What would I expect?' and observes her answer. It is similar to the kind of self-experiment: 'Do I find such and such a sentence grammatical?' We discover what we think is possible: 'Do we think objects can be strapped together [as in Galileo's falling bodies]? Yes, we do. Do we think objects fall with radical discontinuities? No, . . .' These seem to draw on Machian instinctive, unarticulated experiences. A thought experiment, she says, can bring this out, but a Norton-style argument will not.

> . . . thought experiments can provide a fulcrum for the reorganization of conceptual commitments; this explains the way in which they can provide us with novel *information* without empirical input. And by bringing the reader to perform experiments in thought, thought experiments can lead us to reject shaky (and ultimately false) theoretical commitments in light of newly systematized but previously inarticulable *knowledge* about the way the world is.
>
> (1998, 415)

In some respects Kuhn's view and Gendler's are like my own. We all think much is learned by thought experiments and that they cannot be eliminated in favour of something empirically innocent (i.e., an argument *à la* Norton). The big difference between us is this: Kuhn and Gendler think we learn about our conceptual scheme, and only derivatively about the world, while I think we learn about the world, and only secondarily about our conceptual scheme.

Mental models

A mental model is something we make in the mind. They simulate reality and we can manipulate them. To illustrate, here is a little puzzle. A three volume edition of *War and Peace* is on a book shelf, arranged in the usual way. Each volume has 500 pages. A bookworm starts at page 1 of volume 1 and eats in a direct line through to the last page in volume 3. How many pages did the bookworm eat through?

You might have thought like this: 3 volumes × 500 pages each = 1,500 pages in total. Now I tell you that is wrong; try again, this time think about how the books are arranged. You should picture in your mind the three books on the shelf and notice where the first page of volume I and the last page of vol III are located. If you do, you will quickly see the right answer: the worm eats through 500 pages. When you pictured the three volumes in your imagination, you created a *mental model* and you got the correct answer by inspecting it.

Nancy Nersessian (1992, 1993, 2007) and Nenad Miščević (1992, 2007) both see thought experiments as best understood in terms of mental models. (See Johnson-Laird 1983 for a full account of the relevant psychology.)

> Thought experimenting is a principal means through which scientists change their conceptual structures. I propose that thought experimenting is a form of "simulative model-based reasoning". That is, thought experimenters reason by manipulating mental models of the situation depicted in the thought experimental narrative.
>
> (Nersessian 1993, 292)

Mental models, it would seem, explain two important things that are relevant to thought experiments. First, how it is possible to learn about the world just by thinking, and second, the ease and speed of the inference.

> One value of having a mental-models form of representation is that it can do considerable inferential work without the person having to actually compute inferences and can narrow the scope of possible inferences. For example, moving an object immediately changes all of its spatial relationships and makes only specific ones possible.
>
> (Nersessian 1992, 9)

In more detail,

> Briefly, my hypothesis is that what distinguishes thought experiments from logical arguments and other forms of propositional reasoning is that reasoning by means of a thought experiment involves constructing and making inferences from a mental simulation. This is what makes a thought experiment *both* 'thought' and 'experimental'. The original thought experiment is the construction of a dynamical model in the mind by the scientist who imagines a sequence of events and processes and infers outcomes. She then constructs a narrative to describe the setting and sequence in order to communicate the experiment to others, i.e., to get them to construct and run the corresponding simulation and presumably obtain the same outcomes.
>
> (Nersessian 1993, 292)

This is a non-linguistic and non-argumentative process.

> In comprehending a thought-experimental narrative we make use of cognitive structures and operations in common use that have been investigated in some detail by cognitive psychologists. Reading, comprehending, and thinking about stories would seem to epitomize thinking with language. Yet, there is a significant body of cognitive research that supports the hypothesis that the inferences subjects make are derived from constructing and manipulating a mental model of the situation depicted by the narrative, rather than by applying rules of inference to a system of propositions representing the content of the text.
>
> (1993, 293)

I am very happy to agree with the non-linguistic and non-argument claims. More generally, I think the mental models approach does considerable justice to many thought experiments. There may be an in principle problem that I will only mention, but not pursue. One view of the mind, known as computationalism, takes thinking to be a kind of computation. If this the right account, then there simply are no such things as mental models. All thinking is by means of propositions.

For the sake of the argument, let us suppose that mental models are legitimate things and confine our worries to whether they can account for *all* thought experiments. For instance, we can manipulate a physical model, but we don't really manipulate a mental model (e.g., we don't literally rotate a mental model in our heads). Instead, we represent ourselves as manipulating real objects. So, how do we know that this is the right way to represent our manipulations? We may not be discovering things about the model, but rather imposing things on the model.

Consider another example. Start with a male ape. Next, add some female apes. Question: how does the male behave? We might 'discover' that he is aggressive and dominates the females. But this is very likely something we added to the mental model, not something we discovered in it. No doubt, it comes from the background beliefs we have, but then we need to know when those beliefs are right or not. The mental model does not tell us, yet that is the very thing we most want to know. Similar examples could probably be cited in quantum mechanics or in relativity theory. This is not an objection to the psychological claim that we think with mental models. It's a problem about their correctness in drawing significant inferences, as we often do in thought experiments. The problem seems most obvious in the case of moral thought experiments.

Nersessian and Miščević read off the results of a thought experiment from the mental model. They are describing something. Moral thought experiments often result in norms, e.g., it is morally permissible (or impermissible) to do X. But norms in a mental model are no more visible than they are in everyday life.

Miščević (2007) has posed an interesting challenge for me or anyone else who takes intuitions seriously. He discusses Galileo's thought experiment that establishes the equality of all bodies in their rate of fall and gives a mental models account. We start with a mental model of someone (Galileo, if you like), standing atop the Leaning Tower of Pisa, dropping heavy cannon balls and light musket balls. This is straightforward. Next, we imaging attaching a musket ball to a cannon ball and dropping this composite object, as well as the other two. Again, this is straightforward. At some point we bring in a background belief that heavy things fall faster than light things. This is well confirmed by everyday experience (and it's Aristotle's explicit theory). We apply this to the mental model and we see the cannon ball fall faster than the musket ball. But the composite object presents a problem. How will it move? We cannot merely read its motion off the mental model. Since it's heavier than the cannon ball alone, it should fall faster. But, since it's composed of a heavy and a light object, its speed should be between the speed of either, which means it will be slower than the cannon ball alone. The mental model could show us one outcome or the other, but it cannot show us the contradiction. Rather, thanks to some reasoning that is independent of the mental model, we see that there can be no mental model of this contradictory situation. But we also see that the paradox can be resolved by having all bodies fall at the same rate. We can then construct a mental model of this.

It would seem then that we construct two mental models. One sets up the situation. But as we try to fill in the details, we find we can't do it consistently. The second mental model comes after some crucial independent work is done. There is a step in between that is not part of any mental model. I would say it has something to do with seeing into Plato's heaven, that is, with intuition. Others have different accounts. In any case, the assumptions that went into the first mental model are not sufficient to yield the second. Schematically, I will put it this way: $MM_1 + X \Rightarrow MM_2$. For me, X is an intuition; for Norton X is perhaps a logical derivation. A champion of mental models needs a single mental model to cover the whole thought experiment. Failing that, it perhaps could provide some account of sequences of mental models that does not beg the question against rival views. That is, it will need to explain things without X at all, or with an X that is a natural part of a mental models account. So, I'll return the challenge to Miščević : What is X?

What do we see?

What do we see in a thought experiment? Special relativity provides an example that is interesting and instructive. Surprisingly, it is not well known.

Ask almost anyone familiar with basic special relativity about the *appearance* of a rapidly moving object and you will get the answer: it looks contracted. This is the natural answer to give, based as it is on the Lorentz contraction formula: if an object has rest length *L*, then it's length in another frame in which it is moving at velocity *v* is

$$L' = L \times \sqrt{1 - v^2 / c^2}$$

However, this answer is wrong. The mistake is in thinking it *is* contracted, so it must *look* contracted. In fact, it looks rotated. This is a surprising fact, not well known even among physicists. The reason for its unexpected appearance is fairly simple. Imagine a cube moving past at high velocity. The side facing us will be Lorentz contracted, just as the formula stipulates. It is what happens at the back of the cube that matters to the appearance. A photon emitted from the far edge of the back and heading toward our eye will take some time to cross the back; another photon emitted from the near side of the back at the same time as the one from the far side has reached the place were the edge was earlier is emitted, but from a location a bit ahead. They now hit our eye simultaneously. And so do several other photon emitted from various locations at various times, all coordinated to reach our eye together. We thus see the whole back. Depending on the relative velocity of the cube, it will look rotated up to 90 degrees (in the limit of *v* = *c*). (For a exposition, see Weisskopf 1960.)

For a moment, put rotation out of your thoughts and hold to the normal assumption that a rapidly moving object both looks and is contracted. Let's now consider a nice puzzle from special relativity, one that is often taught in standard textbooks.

A car and a garage have identical rest lengths of 4 metres each. The car is approaching the garage at a very high velocity, near the speed of light. The garage owner says the car can easily fit inside the garage, since it will be Lorentz contracted, and hence, much shorter than its rest length of 4 metres. The car owner says the car will not fit inside the garage, since the garage is Lorentz contracted, and hence much shorter than the 4 metres of the car. Both seem to be reasoning correctly and all frames are legitimate, so we are faced with a paradox. How do we resolve it?

The standard resolution—which is certainly correct—is that the relativity of simultaneity plays a role here. In special relativity the order of events can vary from frame to frame. In this case, the garage owner (correctly) says that the rear bumper of the car entered the garage before the front bumper broke through the rear wall. The car owner says (correctly) that the front bumper went through the rear wall before the rear bumper entered the garage.

Pedagogically, this is a wonderful example. Understanding it helps one to understand the central role of the relativity of simultaneity, perhaps the most fundamental feature of special relativity.

But how does one see the paradoxical situation, at least initially? That's easy. In our mind's eye or in a diagram on the textbook page we see the shortened car, if we imagine ourselves in the garage frame, or we see the shortened garage, if we are in the car frame. But in light of what I said above about the appearance of a rapidly moving object, this is not what we would actually see at all. Instead of a shortened car that will easily fit inside the garage, we see a rotated car that will broadside the garage. And similarly, instead of seeing a shortened garage that the car cannot fit within, we see a rotated garage coming at the car, a garage whose door is turned away from the car. Rather than see the car about enter the garage door, we see the car and garage about to hit each other on the side. That is how things would appear if this thought experiment were carried out in reality.

The example is intrinsically interesting, but I think we can draw some lessons from it. The first and most important, is this: *In a thought experiment we see things as they are, not as they would actually appear.* Often appearance and reality go together, but not in all cases, as this example shows. This fact, I think, has implications for various accounts of thought experiments.

A popular account of thought experiments claims they are continuous with real experiments. Roy Sorensen has especially promoted this view in his very important and influential book, *Thought Experiments* (1991). Once we allow that there is a thought experiment in which reality is what is seen rather than appearance, then the continuity thesis, as it is known, must be abandoned.

Similarly, there is a stipulation often laid down which must be abandoned. It is the stipulation that a good thought experiment should in all respects be as close as possible to what we would actually observe. But adding the appearance of rotation would actually get in the way and ruin this thought experiment. Of course, this does not always happen. Often adding information to make a thought experiment more like a real one, is simply useless, but not harmful, as, for instance, saying what colour Galileo's cannon and musket balls are.

Platonists, of course, delight in any appearance–reality distinction, so the car–garage example is a happy one for us. As I noted, I think it hurts the claim that thought experiments and real experiments are on a continuum. As for other accounts, it is hard to say. If a good mental model should be as close to what we would actually experience, then the mental models account of thought experiments will be hurt, as well. Norton's argument view is, I think, untouched by the relativistic appearance example. However, the example is interesting in its own right, regardless of its bearing on any of the various views of thought experiments.

A *priori* but false?

Throughout this book I've insisted that *a priori* reasoning is fallible. The conclusion of the Galileo thought experiment has recently been challenged and the EPR conclusion is almost certainly wrong, given the Bell results. Let's suppose for the sake of the argument that both of these wonderful thought experiments actually do result in a false conclusion. How then can I reasonably claim that the thought experimenter grasps the relevant abstract entities or sees the particular law of nature involved? In such cases there can be no such entity to be grasped, no such law of nature to be seen. Perhaps Kuhn is right in thinking there is no more to the world than is created by our conceptual scheme, and that thought experiments really tell us about those, not about a paradigm-independent world.

Frankly, I have no idea how to answer this question, which is one of the most important in all philosophy of science. It is not a problem which is peculiar to my view—it is really the problem of verisimilitude which has been the bane of philosophy of science for years. How is it that a (physical) world that contains no phlogiston, caloric, or aether can somehow be responsible for bringing about the phlogiston, caloric, and aether theories? Though we often now make fun of such theories, they were actually successful to some degree in their day and were believed by reasonable people. (Maxwell once said that the aether theory was the best confirmed in all science.) The physical world somehow or other contributed to the production of these rational, but false, beliefs. It seems fair to consider them as steps in the series of better and better theories about the world. They are false; they do not 'cut reality at its joints'; but they are not totally disconnected from the world either.

In just the same sense, I would claim, Galileo's thought experiment and that of EPR hook on to abstract reality even though they may not cut it at its joints either. And even though the thought experimenter does not perceive things clearly, the abstract realm nevertheless contributes to or causes the belief. *A priori* fallibility presents no more problems than empirical fallibility does. Mistakes are a mystery anywhere they occur—for that matter, so is getting it right.

Possible worlds reasoning?

Recent years have seen great interest in what are called 'possible worlds'. (I have used the idea loosely in a number of places already in this book.) Possible worlds are a semantic device which has been put to great use in modal logic (which deals with *necessity, contingency,* and other modal notions) and the analysis of counterfactuals (i.e., conditional sentences of the form: If P *were* to be the case then Q *would* be the case). Success in these realms naturally leads one to suspect that possible worlds will shed some light on thought experiments as well. There are no frictionless planes in *our* world, so let's just consider a possible world in which there are and then see how Stevin's chain device behaves. No one here can run at the speed of light, so let's consider a possible world in which people can, then ask what they see.

At this vague level, talk of possible worlds is harmless. Indeed, it is even heuristically useful. But I am not too sanguine about the utility of possible worlds when we push for details.

Possible worlds are ways things could be. They can be quite fantastic: filled with talking horses, immortal butterflies, and objects which fall at different rates due to their different colours. But there is one thing any possible world must be: consistent. This—perhaps surprisingly—is what makes them inadequate for the analysis of a great deal of scientific reasoning. The simple fact is that many thought-experimental situations, like the scientific theories they deal with, are outright inconsistent.

Let us consider Maxwell's demon for a moment and let us demand to know how the demon knows the whereabouts of the molecules it is sorting.[10] Either we have an ignorant or an omniscient demon. (A) If the demon is initially ignorant then it has to gather information about the position and velocity of the molecules. Perhaps it does so with a flashlight, bouncing photons off the various molecules. In this case the demon does work, arguably more than is gained in having the heat transferred from the cold to the hot body. In such a case the thought experiment is pointless. Household refrigerators can already do that. (B) If Maxwell's demon is omniscient—like Laplace's demon, it already knows the initial positions and momenta of all the molecules—then it is some kind of supernatural being, a kind of dynamical system which is not itself subject to thermodynamical constraints. The demon seems to violate the very laws it is meant to illustrate. I think it is clear that Maxwell intended the second, omniscient version. Though the whole situation is contradictory, the thought experiment works, nevertheless. The violation of the second law of thermodynamics is not itself contradictory, but it may take a contradictory situation to get us to see its possibility.

This is true, not only of many relatively visualizable thought experiments but of much scientific reasoning of the more usual discursive sort as well. For example, anyone drawing inferences from quantum electrodynamics is (arguably) reasoning from inconsistent premises.[11] Much scientific reasoning—and good scientific reasoning at that—is in the spirit of the White Queen:

> ALICE: There's no use trying, one *can't* believe impossible things.
> WHITE QUEEN: I daresay you haven't had much practice.
> When I was your age, I always did it for half-an-hour a day.
> Why, sometimes I've believed as many as six impossible things before breakfast.
>
> (Lewis Carroll, *Through the Looking Glass*)

Possible worlds were introduced to clarify semantical notions such as meaning and truth. In trying to understand thought experiments we have been concerned with epistemology, so we need not be too surprised that this logical apparatus is none too helpful. Why, for instance, should we think that reasoning about any possible world will tell us something important about our own? We must first be informed that the world we're thinking about is similar to ours in the relevant respects. And this, of course, is the one thing we usually don't know and are trying very hard to find out.

But even when confined to purely semantical topics in the philosophy of science, possible worlds have been less than impressive. David Lewis (1986), for example, suggests that possible worlds might give an adequate analysis of the notion of *verisimilitude*. Prior to Lewis, no account of one theory being closer to the truth than another has worked, and I doubt that Lewis's fares any better. Some possible worlds, he claims, are closer to, or more similar to, the actual world than others. Lewis uses this primitive notion of closeness of worlds to analyse T1 *is closer to the truth than* T2 as follows: T1 is true in W1 and T2 is true in W2 and W1 is closer to the actual world than W2.

Such an analysis may do considerable justice to transitions in the history of science like that from Newtonian to Einsteinian physics. But the idea of verisimilitude must include (inconsistent) QED being closer to the truth than, say, Franklin's (consistent) two-fluid theory of electricity. When it comes to possible worlds, QED is either true nowhere or else is true in the ficticious world (where all contradictions hold), a world that is farther away than all others.[12] So it seems that a possible worlds analysis of truth-likeness can do no justice to this or any other contradictory theory—yet the history of thought would be impoverished without them.[13]

Galileo as a rationalist

Galileo's writings have provoked endless controversy about the respective roles of reason and experience. One of the pole positions has been very well articulated and defended by Alexandre Koyré, the French philosopher and historian of science who sees Galileo as a Platonist, a brilliant rationalist who knew 'how to dispense with [real] experiments . . .' (1968, 75). 'Good physics', says Koyré, 'is made *a priori*' (1968, 88). On the other hand, a very plausible empiricist interpretation of Galileo has been developed in great detail by Stillman Drake (1978), who holds that Galileo was a brilliant experimenter who paid no attention to philosophy.

I suppose great scientists like Galileo must endure our attempts to make them over in our own image.[14] Needless to say, I incline to Koyré's view. Nothing better sums up Galileo's rationalism than passages Galileo himself puts into the mouth of his spokesman Salviati in the *Dialogo* in conjunction with listing some of the problems with the Copernican theory.

> No, Sagredo, my surprise is very different from yours. You wonder that there are so few followers of the Pythagorean opinion, whereas I am astonished that there have been any up to this day who have embraced and followed it. Nor can I ever sufficiently admire the outstanding acumen of those who have taken hold of this opinion and accepted it as true; they have through sheer force of intellect done such violence to their own senses as to prefer what reason told them over that which sensible experience plainly showed them to the contrary.
>
> (*Dialogo*, 327f.)

> These are the difficulties which make me wonder at Aristarchus and Copernicus. They could not have helped noticing them, without being able to resolve them; nevertheless they were confident of that which reason told them must be so in the light of many other observations. Thus they confidently affirmed that the structure of the universe could have no other form than that which they had described.
>
> (*Dialogo*, 335)

we may see that with reason as his guide [Copernicus] resolutely continued to affirm what sensible experience seemed to contradict. I cannot get over my amazement that he was constantly willing to persist in saying that Venus might go around the sun and be more than six times as far from us at one time as at another, and still look always equal, when it should have appeared forty times larger.

(*Dialogo*, 339)

The status of the theory

It is often interesting to ask self-reference-type questions about a theory. Someone conjectures that things have such and such a property. Does the conjecture itself have this property? This kind of question is sometimes used as a club to bash the likes of scepticism and Marxism: The sceptic says no belief is justified; therefore (comes the reply), scepticism is not justified either. Marx says political beliefs are ideology reflecting class interests determined by the mode of production; therefore (comes the reply), Marxism is merely ideology determined by material factors as well. Such quick retorts are usually unfair; they are certainly unfair in the cases of scepticism and Marxism.

Here is a more interesting case. Kripke (1980) and Putnam (1975c) have both argued that natural kinds have some of their properties necessarily. Thus, if water is H_2O then it is necessarily H_2O. A proposition is a necessary truth if and only if it is true in every possible world. Now let us ask: Is the Kripke-Putnam theory of necessary truths itself a necessary truth? The answer is yes; if it is true at all, then it is necessarily true.

The argument is simple. Suppose the theory is true, but not necessarily true. Since the theory is not necessarily true, there is a possible world where it is false. In order for the theory to be false, there must be a world where there is a counter-example to one of its necessary truths, say, that necessarily water is H_2O. But this would be a possible world where water is not H_2O, which contradicts the assumption that the theory is true. Consequently, the Kripke-Putnam theory, if true at all, is necessarily true. So this philosophical theory must be true everywhere or nowhere.

What about the account of thought experiments which has just been outlined? What is its status? I have argued that some of our knowledge of nature is *a priori*. Is the same true of my own theory? Let me start by contrasting my own account with the epistemology of Descartes.

As is well known, Descartes thought we have *a priori* knowledge of the world. Moreover, this belief itself was taken by him to be *a priori.* Here is why: he started with universal doubt; but he could not doubt that he was doubting; *cogito, ergo sum,* as he put it. Next, Descartes had a proof of God's existence and he was certain God wouldn't fool him, since deception is an imperfection; thus he could trust his clear and distinct ideas. So, not only is our knowledge of nature *a priori,* according to Descartes, but how he arrived at this epistemological theory is also *a priori.*

Here is where I part company with most traditional rationalists. Like Descartes, I hold that (some of) our knowledge of nature is *a priori;* but unlike Descartes, my belief that this is so is a *conjecture.* I have no *a priori* argument for it. Rather I am hypothesizing it in order to explain a peculiar phenomenon, namely, successful thought-experimental reasoning in the sciences. I propose that there are independently existing laws of nature and that we have some sort of capacity for grasping them as the best explanation of that phenomenon. It is my firm conviction that one of the aims of epistemology—perhaps the chief aim—is to explain the history of science. My account of thought experiments is offered as an explanation of one important aspect of science and should be evaluated in those terms.

The grounds for evaluating any conjecture have to do with its explanatory power, its ability to unify diverse phenomena, and its success in making novel predictions. The last of these doesn't seem appropriate in evaluating a philosophical theory, which is just as well since I have no novel predictions to offer, anyway. But the other two features should be kept in mind when looking at the following chapters.

6 The development of inertial motion*

The history of mechanics has been extensively investigated in a number of historical works, including classic books such as Mach (1960), Dugas (1988), and Clagett (1961), and scores of articles in various academic journals. The full story from the Greeks and medievals through the Scientific Revolution to the modern era is long and complex. It is also incomplete. Studies to date have been admirably thorough in putting empirical discoveries into proper perspective and in making clear the importance of mathematical innovations. But there has been surprisingly little regard for the role of thought experiments in the development of mechanics. Of course, many historians have rightly stressed the importance of reconceptualization in the history of science and further noted that this process is something different from the observation of new facts. But the focus has not been on thought experiments as the driving force behind such conceptual developments. In neither kinematics nor dynamics have they received their historical due in their ancient, medieval, classical (Newtonian) or modern (Relativity and Quantum Theory) forms, in spite of their obvious importance in the development of each.

Perhaps this can be rectified to some extent, but there is no hope of being exhaustive. Only one theme—thought experiments in connection with inertia and relative motion—will be the focus of this chapter. A full study of even a single period in the history of mechanics would, of course, include hard won empirical data, the development of instruments such as the telescope, and relevant mathematical innovations, as well as thought experiments. The limited aim here is to present a brief history of one aspect of mechanics via thought experiments. And, narrow though this brief study may be, it will be further truncated. It should start with the Greeks and finish with relativity, but after some Greek preliminaries, it will be confined to the late medieval period and the early Scientific Revolution, years that were crucial in the development of classical mechanics and the idea of inertial motion.

* This chapter is based on an article written with Mark Shumelda.

The ancient Greek outlook

The Presocratics

As with so many things, there is no better place to start than with the Greeks. Early thinking about space was confused and confusing—but often ingenious. According to Aristotle (on whom we rely for much of our knowledge of the pre-Socratics),

> Leucippus and his associate Democritus hold that the elements are the full and the void; they call them being and not-being, respectively. Being is full and solid, not-being is void and rare. Since the void exists no less than body, it follows that not-being exists no less than being. The two together are the Material causes of existing things.
>
> (*Metaphysics*, A4, 985b4)

One can immediately see why the void was so controversial for the Greeks. The very expression of the idea is problematic. The void is nothing, it is non-being. So, one might naturally conclude that, since the void is nothing at all, it does not and cannot exist. That's exactly what many, including Parmenides, did conclude. The idea of the void was dismissed as illogical nonsense. Yet, for many others, the void was necessary, for how otherwise could things move if they did not have empty room to move into?

Parmenides and his followers flourished at about the same time as the early Atomists (fifth century BCE). The Parmenidean objection was simple: the void is nothing and nothing cannot exist as a thing. Whatever does exist must be a continuous, undifferentiated whole. Change requires things moving about, but this would be impossible, since there is no place for anything to move into. Consequently, claims Parmenides, the universe is unchanging, and the appearance of change and difference is an utter illusion—reality is wildly different than it seems.

Zeno's contribution to the Parmenidean outlook was generated by several famous thought experiments. They took the form of paradoxes in which he would show that the common-sense supposition of motion leads to absurdities. In one of these paradoxes Zeno argues that it is impossible to traverse a stadium. Before getting to the other side, one must get to the mid-point. But before reaching the mid-point, one must get one quarter of the way, and so on *ad infinitum*. Thus, to traverse the stadium, one would have to pass through infinitely many distances in a finite time, and this was taken to be impossible.

Another of Zeno's paradoxes shows that Achilles cannot catch the tortoise, if the tortoise has a head start. It matters not how fast Achilles can run, how slow the tortoise moves, or how short the gap between them. If Achilles runs the distance to where the tortoise started in a given time interval, then the tortoise has moved ahead a bit. Achilles must now run that second distance, but during that time the tortoise has moved ahead a bit more. Though the gap between them is getting smaller, there are infinitely many such gaps to be covered, and this will be impossible for Achilles to achieve in a finite time.

The conclusion that Zeno wants us to draw from his paradoxes is that our ordinary notions of space, time, and motion are confused and contradictory. His thought experiments support Parmenides by showing that any contrary view is absurd—even though it is unshakable common sense. Appearance and reality sharply diverge.

As is so often the case in fundamental differences of opinion, one side's *modus ponens* is the other side's *modus tollens*. Parmenides and his followers thought the void is absurd, so they drew the conclusion that motion is impossible. The Atomists turned this around, claiming that motion is manifestly possible, and so they concluded that the void must exist, after all. What the two sides shared is the common premiss: *motion requires the void*. Greek thought about space and motion seemed at an impasse, at least until Aristotle (384–322 BCE).

Aristotle

Space and time are part of Aristotle's general theory of motion and change. Not only was he concerned with how an object changes from one spatial location to another, Aristotle was also interested in how an acorn changes into an oak tree. He was, in short, concerned with motion or change in a very general sense, and he made use of the concepts of *potentiality* and *actuality* in his explanation. There are a number of important ingredients in Aristotle's account.

First, *the non-existence of the infinite*. It's a commonplace to say there are infinitely many points on a line. The great success of analytic geometry (which puts the real numbers into one–one correspondence with the points on a line) no doubt accounts for this widespread view. It has become unquestioned common sense. But is it true? Aristotle certainly thought not. There are no points on a line (or on a plane or in space), except those we make. The line, according to Aristotle, is a true continuum. It is infinitely divisible in the sense that it can be divided again and again without limit. If we make a cut in a line, say, with a compass, then we construct a point. We may do this repeatedly and without limit; that is, we never arrive at a situation where we cannot do it again. But at any given stage in the process of constructing points, we have made only finitely many cuts in the line, never a so-called *actual (or completed) infinity*. The only infinity for Aristotle is the *potential infinity*.[1]

With this account of the infinite in mind, Aristotle could tackle Zeno's paradoxes. The crucial premiss in Zeno's arguments is the assumption that space consists in an actual infinity of points and time in an actual infinity of moments. Aristotle's theory of the potential infinite flatly denies this. So, Zeno's arguments can't even get started, since the necessary preconditions for the thought experiments do not exist.

Second, *natural place and natural motion*. Nature, according to Aristotle, is the cause of motion and change. *Place* is defined as 'the innermost motionless boundary of that which it contains' (*Physics*, 212a20). The place of the wine is in the bottle; the place of the boat is in the water that immediately surrounds it. This is a relativistic view of spatial location, especially when motion is considered. But there is a hierarchy of locations. The place of the moving boat is in the water; the place of the flowing water is in the river banks; the place of the river banks is . . ., and so on. In the ultimate scheme of things there is the stationary centre of the universe, the ultimate frame of reference.

These ideas are used in Aristotle's elaborate and far-reaching doctrine of *natural place*. The universe is finite in size (since space cannot be actually infinite), but it is not in any place (since there is no container). There is a centre of the universe and that is where the earth is. There is an objective 'up' and 'down', defined in terms of radial lines from the centre of the universe. Nature directs things to their natural place. A rock falls toward the centre of the earth, not because of the action of the earth, but because the rock strives to get to its natural place, the centre of the universe. If we could somehow move the earth away a few thousand kilometres and then drop the rock, we would not see it fall to the earth, but rather it would move toward the centre of the universe. Similarly, there is a natural place for water which is just above the natural place for earth, air above that, and fire, yet again is naturally above air. When removed from their natural places, they strive to return to them.

Interestingly, this was the basis of a thought experiment to show there are no worlds like ours elsewhere in the universe. (The Atomists had thought there must be.) Suppose there were such a world with trees, mountains, lakes, people who live in houses, and so on. But the rocks on the mountain and the bricks in the house would naturally move to their natural place, the centre of the universe, which is not the centre of that world. Our world, earth, is the only world that is at the centre of the universe. All others are hopelessly unstable and would quickly disintegrate.

Just as Aristotle appeals to 'naturalness' in his definition of place, so he has a corresponding idea of motion. Material objects are more or less defined by Aristotle in terms of their natural motion. This may sound foreign to modern ears, but we should remind ourselves that properties like this are present in current physics. While electrons are not defined in terms of velocities (any below the velocity of light being allowed), photons, for example, are characterized by the fact that in every inertial reference frame they always move at velocity c, a motion which is 'natural' for them. In Newtonian mechanics, as long as no force is acting on it, a body moves in a straight line at a constant speed. This motion, inertial motion, is natural for any body. And in general relativity motion along a geodesic is natural motion. Some of the great re-conceptualizations in the history of science have involved something previously thought to be non-natural (and in need of explaining), and making it natural and capable of explaining other things. Inertial motion is a prime example.

Aristotle's distinction between natural and non-natural motions (which he called violent) is crucial. Even though we differ greatly in detail with Aristotle, we do adopt something similar in distinguishing kinematics (free motion) from dynamics (motion under a force). Though conceptualized in a variety of different ways, the distinction has been remarkably stable, transcending particular theories.

There are four elements in the Aristotelian universe—earth, water, air, and fire—and their natural places are in concentric rings around the centre of the universe. Beyond these elements is a fifth, the aether. The natural motion for the four elements is along radial lines through the centre of the universe, toward its natural place. Natural motion for the aether is circular, again around the centre of the universe.

Why is aether's natural motion circular? Aristotle's answer would have rung true to any intelligent listener until the seventeenth century. Circular motion, he said, is perfect. Not only is it perfect, but circular motion is unique in leaving the perfect celestial realm perfectly unchanged. It rotates around the earth (remember, that's the centre of the universe), as a rigid, unchanging whole.

Third, *non-existence of the void*. Aristotle denied the existence of the void, empty space. The universe is full; there is stuff of one sort or another everywhere. He offered several arguments in various places against a void or a true vacuum. One runs as follows: there is no difference of any sort among the various regions of a void; it is completely homogeneous. But a rock, for example, moves to its natural place, and in order for this to happen, one place must be objectively different from another. Since one place is not different from another within a void, motion in a void must be problematic, if not impossible. A rock in a void would be completely disoriented; it wouldn't know which way is down, and so wouldn't know which way to move, or whether to move at all. Aristotelian place has a kind of causal power, or at least the power to tell an object how to move; the void has no such power.

Aristotle had a very big problem to worry about. Remember the dispute between the Atomists and Parmenides. Both agreed that the void is necessary for motion, but Parmenides thought the void absurd, so motion must be impossible, while the Atomists thought motion is obvious, so, the void must exist. Aristotle wanted it both ways: there is motion and it happens in a full space.

It seemed, of course, that if everywhere were to be filled with stuff, then there would be no room for anything to move. Aristotle countered that motion is possible in a plenum so long as one body moves aside for another. The idea is really quite simple. Think of a fish swimming through the water which is displaced in front and moves in behind as the fish swims along. We'll come back to this idea momentarily.

Fourth, *natural vs violent motion*. When a rock is thrown up, it is certainly moving, but not naturally, in Aristotle's sense. Such motion was called 'violent motion' and making sense of it was horribly difficult for Aristotle and his followers right through the Middle Ages. When a rock or spear leaves our hands, why does it persist for a while in its unnatural motion and why does natural motion eventually overtake it? As we just mentioned, the account that Aristotle seemed to favour claimed that the air, which got out of the way of the moving object, rushed around to the back (to prevent a vacuum) and this gave the object a push. Thus the object would be kept in violent motion even after it left our hands. The process is called *antiperistasis*.

Finally, we should point out, too, that Aristotle's universe is finite in spatial extent. The Atomists, both early and late, took the void to be infinite. Lucretius, in his *De rerum natura*, offered a clever thought experiment to prove it (I, 951–968). It was described earlier, but it won't hurt to describe it again. Suppose space is only finite. Then, if we come to the edge or boundary, we could throw a javelin. If it flies past the supposed edge, then it is not the edge after all. On the other hand, if the javelin bounces back, then there must be something beyond the edge that makes it bounce back. So, once again, the supposed edge is not really an edge after all. Thus, space has no edge; it is infinite.

Summary

The point of this brief section on Greek thinking about space and motion has been to set the stage for the medieval and early modern period.

Both the difficulties that were faced and the conceptual tools that were employed during these eras are the legacy of the Greeks. The medievals were in the grip of Aristotle and the Aristotelian framework. To summarize, the chief ingredients of the medieval framework inherited from Aristotle include:

- The objects we see around us consist of a mix of four elements, earth, water, air, and fire. The heavens are made entirely of aether (the fifth element or quintessence).
- Each element has a natural place and strives to move to it.
- Natural motion for the four elements is in radial lines through the centre of the universe. Natural motion for aether is circular.
- Violent motion is explained by antiperistasis—whereby the medium through which an object travels is actively responsible for keeping the object in motion.
- The universe is full; there is no vacuum (no void, no empty space).
- Everything that moves unnaturally requires a mover. There is nothing akin to conservation of energy or momentum; things naturally come to rest in their natural place.
- God (the unmoved mover or movers), is constantly injecting motion into the universe.

I will expand on some of these points below. For now I stress that there is nothing here in the way of a concept of relative motion or inertia. I will take a liberal view of what inertia is, but to be a reasonable approximation to our modern concept, it should include the possibility of motion as well as rest as a 'natural state' of being.

Challenges to Aristotelian dynamics

Natural vs violent motion

Why does a rock thrown horizontally continue in its horizontal motion after it has left the thrower's hand? This was a serious puzzle, since its motion is not natural. It should, in the Aristotelian view, move down, toward its natural place at the centre of the universe. Such non-natural motion, as earlier noted, is called 'violent', and its explanation was one of the central problems in Aristotle's physics. Why does violent motion continue after an object has been detached from the source of its motion?

According to Aristotelian dynamics, every motion presupposes a mover. In order to satisfy this requirement for the case of detached motion, Aristotle explains that the original cause of motion imparts a moving power not only to the projectile itself, but also to the medium surrounding the moving object:

> If it is true that everything that is moved . . . is moved by something, how comes it that some things are moved continuously, though that which has caused them to move is no longer in contact with them, as, for instance, things thrown? . . . We must, therefore, hold that the original movement gives the power of causing motion to air, or water, or anything else which is naturally adapted for being a movent [i.e., mover] as well as for being moved.
>
> (*Physics* VIII.10.266b,27–267a.20, Clagett 1961, 506)

Aristotle's theory of antiperistasis supposes that while the medium remains stationary during the object's movement through it, it nevertheless 'simultaneously [along with the object] receives the power or force to act as a movent' (1961, 507). So the particles of air closest to the original mover, for example, receive the full amount of motive force—some of which they pass onto their neighbouring particles, which then pass a reduced amount of motive force onto yet further particles, and so on. Each successive 'wave' of air actively pushes the projectile, thanks to the 'the power to cause motion' (1961, 506) which it receives from its neighbouring wave of air. Resistance causes this motive force to be gradually expended. The motion of the object 'finally ceases when the member of the series [of particles in the medium] immediately preceding no longer makes the next a movent but only causes it to be moved' (1961, 507).

It is important to note that in Aristotelian dynamics, resistance plays a fundamental role. Without it, no motion could take place.

> Without resistance, motion does not take place, and the absence of resistance is the cause of why a simple body is not moved in a vacuum. Because a simple body does not have intrinsic resistance, and there is no extrinsic resistance in a vacuum, and [further] there is no resistance except intrinsic or extrinsic [resistance], it follows that a simple body does not have resistance, and consequently is not moved in a vacuum. Because without resistance motion does not take place—except one that is infinitely fast—it follows that a pure, simple body is not mobile in a vacuum unless it should be mobile infinitely fast; because if this were the case, it would traverse space immediately, which is impossible.
>
> (*An Anonymous Treatment of Peripatetic Dynamics* 460, Clagett 1961, 451)

The dynamics of antiperistasis (in violent motion the medium pushes the moving body) remained dominant well into the high Middle Ages, but was plagued with conceptual and empirical problems. Some of the most effective and influential criticisms came in the form of thought experiments, especially those advanced by Jean Buridan.

Violent motion

Natural philosophy in the fourteenth century was marked by a period of transition. Thinkers such as Jean Buridan (ca.1300–ca.1361) and his student, Nicole Oresme (1323–1382), raised serious criticisms against the prevailing Aristotelian theory of dynamics. In its stead they proposed the theory of impetus, which claims that the source of motion is not to be found in the medium (i.e., moving to natural place or the medium pushing) but rather in the moving object itself. In some ways, this development prefigures the discovery of the laws of inertia[2] by Galileo Galilei (1564–1642) and René Descartes (1596–1650) two centuries later. What draws Buridan, Oresme, and Galileo together is their shared use of thought experiments, many of which are among the most brilliant ever conceived.

Each of the thought experiments to be discussed identifies some instance of detached (i.e., violent) motion (whether projectile or circular) which is presumably derived from experience and yet fails to be explained by the tenets of Aristotelian dynamics. In each case, the reader is supposed to conclude that Aristotle's theory is seriously deficient.

Buridan's commentary on Aristotle's *Physics* contains the following devastating thought experiments, some of which are close to reports of actual experience, while others are more hypothetical. They are all directed against Aristotle's theory of antiperistasis.

> A lance having a conical posterior as sharp as its anterior would be moved after projection just as swiftly as it would be without a sharp conical posterior. But surely the air following could not push a sharp end in this way, because the air would be easily divided by the sharpness.
>
> (Buridan, *Questions on the Physics of Aristotle* VIII.12.2, Clagett 1961, 533)

> A ship drawn swiftly in the river even against the flow of the river, after the drawing has ceased, cannot be stopped quickly, but continues to move for a long time. And yet a sailor on deck does not feel any air from behind pushing him. He feels only the air from the front resisting him.
>
> (1961, 53)

Again, suppose that the said ship were loaded with grain or wood and a man were situated to the rear of the cargo. Then if the air were of such an impetus that it could push the ship along so strongly, the man would be pressed very violently between that cargo and the air following it. Experience shows this to be false.

(1961, 53)

Or, at least, if the ship were loaded with grain or straw, the air following and pushing would fold over the stalks which were in the rear. This is all false.

(1961, 533)

The moral is clear. In each case, the air cannot be moving the very objects it is purported to move. If it did, we would notice effects plainly contrary to common experience and expectation. Antiperistasis cannot explain these cases of violent motion.

Another thought experiment by Buridan makes use of a very straightforward analogy between launching a projectile together with its surrounding air and pushing the surrounding air by itself.

You could, by pushing your hand, move the adjacent air, if there is nothing in your hand, just as fast or faster than if you were holding in your hand a stone which you wish to project. If, therefore, that air by reason of the velocity of its motion is of a great enough impetus to move the stone swiftly, it seems that if I were to impel air toward you equally as fast, the air ought to push you impetuously and with sensible strength. [Yet] we would not perceive this.

(1961, 534)

With this thought experiment Buridan makes the case against antiperistasis even more devastating. If I throw a big stone at you, it will hurt. But if I throw the same amount of air that is pushing the stone, it seems as nothing to you. This thought experiment has the same experiential flavour as any great experiment, but the conclusion is so obviously absurd, that we feel no need to carry it out.

The structure of these thought experiments is a common one. We set up the conditions and see what happens. But we find that things do not go as Aristotle's theory predicts. The conclusion we draw is that Aristotle is wrong. Many of Galileo's famous thought experiments, as we saw and shall see again, have this form. And some of Einstein's, too, as when we earlier described him as chasing a light beam to see what an electromagnetic wave front would look like. It led to an absurdity. Such negative thought experiments are often very effective in telling us what is false, but they do not tell us what is true. And this is the case with Buridan so far. We can see what is wrong with Aristotle, but not yet what is right. This will continue to be the case with the next few examples. Later we will see thought experiments that play a positive role in coming to a new theory.

Let us now turn to circular motion, which presents a particularly difficult challenge to antiperistasis. In this thought experiment we are to consider a spinning top and a smith's mill (a heavy stone wheel for grinding grain),

> which are moved for a long time and yet do not leave their places. Hence, it is not necessary for the air to follow along to fill up the place of departure of a top of this kind and a smith's mill. So it cannot be said [that the spinning top and smith's mill are moved by the air] in this manner.

> (1961, 533)

There is no need for the air to fill in the places vacated by a rotating object, since those places are filled in by other parts of the object itself. And yet the wheel continues to move. Antiperistasis is gratuitous at best in such cases and downright false at worst.

Another of Buridan's thought experiments that deals with detached motion is particularly intriguing, because, like others it would seem to be do-able, but in fact would be quite challenging to carry out. Nevertheless, the result seems perfectly evident, from merely visualizing the situation.

> If you cut off the air on all sides near the smith's mill by a cloth, the mill does not on this account stop but continues to move for a long time. Therefore it is not moved by the air.

> (1961, 533)

This thought experiment differs from the previous examples insofar as it represents Buridan's attempt to create an idealized or fully 'controlled' experimental scenario. All of the thought experiments encountered thus far argue against the theory that the air (or the medium in general) is responsible for projectile or circular motion. All that Buridan's previous counter-examples accomplish, however, is to demonstrate that the air does not visibly 'push' or 'mutually replace' itself behind a moving object.

The examples, however, are quite realistic and assume that the objects remain immersed in the air. A clever Aristotelian could possibly, therefore, come up with some kind of dynamical theory that still involves the medium—perhaps as a mysterious 'unperceived' or 'unobserved' mover. Buridan's last thought experiment is the only one that can counter this neo-Aristotelian rejoinder. This is because it actually proposes *removing* the air altogether from the scenario—rendering it impossible for the medium to be responsible for sustaining detached motion. Unlike the previous thought experiments, which involve common, everyday occurrences, like throwing lances and observing passing ships, Buridan proposes that we abstract from the ordinary scenario of a 'merely' spinning wheel. Admittedly, it is far from clear what 'cut[ting] off the air on all sides near the smith's mill by a cloth' would entail. Does Buridan intend for us to imagine the wheel spinning in a complete vacuum? Perhaps all he means is that the cloth is to enclose the spinning wheel as tightly as possible without actually touching it. In any event, what is clear is that Buridan constructs this last thought experiment so that the air can play *no role whatsoever* in the motion of the wheel. This move to an idealization removes all doubt as to the failure of Aristotelian dynamics.

Jumping ahead to Galileo, we see the same sort of response to anti-peristasis. In the following thought experiment, Galileo uses an analogy to set up a *reductio ad absurdum* argument against Aristotle's theory of projectile motion.

> SAGREDO: If I were to place two arrows upon that table when a strong wind was blowing, one in the direction of the wind and the other across it, the wind would quickly carry away the latter and leave the former. Now apparently the same ought to happen with two shots from a bow, if Aristotle's doctrine were true, because the one going sideways would be spurred on by a great quantity of air moved by the bowstring—as much as the whole length of the arrow—whereas the other arrow would receive the impulse from only as much air as there is in the tiny circle of its thickness.
>
> SIMPLICIO: I have never seen an arrow shot sideways, but I think it would not go even one-twentieth the distance of one shot point first.
>
> (*Dialogo*, 153)

This thought experiment could just as easily have come from Buridan. And Galileo would feel perfectly at home with any of Buridan's examples. In attacking Aristotle, they are largely interchangeable. But how they responded to the problems with Aristotle, as we shall see, is quite different.

Still on the theme of the air being responsible for motion, Galileo points out another type of problem, this time concerning relative weights of various things and the fact that air itself has the least weight of all.

SALVIATI: [Consider what would happen] if two strings of equal length were suspended . . . with a lead ball attached to the end of one and a cotton ball to the other. Which of these pendulums do you believe would continue to move the longer before stopping vertically?

SIMPLICIO: The lead ball would go back and forth a great many times; the cotton ball, two or three at most.

SALVIATI: If the pendulums have just shown us that the less a moving body partakes of weight, the less apt it is to conserve motion, how can it be that the air, which has no weight at all in air, is the only thing that does conserve the motion acquired?

(*Dialogo*, 152)

Galileo provides us with the beginnings of a positive explanation for why a heavy pendulum swings longer than a light one (and what causes violent motion in general). According to the above analogy, dense, solid objects are capable of 'conserving motion' while the gaseous air is not. Galileo thus posits that whatever might be the source of projectile motion, it is not to be found within the medium through which the object travels, but within the object itself. The thought experiment not only presents a plausible phenomenon with which to refute the rival dynamical theory of Aristotle, but begins to develop a positive account of projectile motion.

Motion of the heavens and earth

Some of Galileo's most important thought experiments on the subject of inertial motion are to be found within the context of his discussion of the diurnal rotation of the earth. In fact, the *Dialogue Concerning the Two Chief World Systems* (*Dialogo*) is chiefly concerned with refuting Aristotle's theory that the earth stands still and that the heavens revolve around it. Galileo's discussion is prefaced somewhat in Oresme's *Le Livre du ciel et du monde*, where the author likewise confronts Ptolemy's geocentric, stationary-earth model of the universe. It turns out that both Ptolemy and Aristotle argue their case by way of thought experiments that involve analogies. Oresme and Galileo thus have their work cut out for them. Each points out the fallacy in his predecessors' analogies and they revise their thought experiments to explicitly incorporate what later became known as the principle of Galilean relativity (i.e., the laws of motion are the same in all reference frames that are moving in a uniform, non-accelerated, manner). The end result, in the words of Galileo's Salviati, is somewhat inconclusive: '[I have claimed] only to show that nothing can be deduced from the experiments offered by [the Aristotelians] as one argument for [the earth's] motionlessness' (*Dialogo*, 154). Because the *local* kinematic behaviour of objects would be unaffected by the uniform rotation of the earth (strictly speaking, since the earth is accelerating by rotating, it cannot be considered an inertial frame), we cannot use thought experiments to settle the question one way or the other. Although Oresme and Galileo fail to provide us with a successful, positive argument for the earth's rotation, by drawing on the principle of Galilean relativity, they implicitly adopt a new dynamical theory: the law of inertia.

Here are two analogy-type thought experiments which claim to prove the immobility of the earth. In the first example, Ptolemy (as presented by Galileo) draws an analogy between a moving ship and the earth.

> If one were in a ship which was moving rapidly towards the east and if one were to fire an arrow straight upwards, it would not fall in the ship, but very far away, towards the west. And similarly, if the earth is moved very quickly in turning from west to east, supposing that one were to throw a stone directly upwards, it would not fall in the place from which it started but far away towards the west; and the contrary indeed happens.
>
> (*Dialogo*, 154)

And, thus, in Ptolemy's view the earth cannot move. In its day, this thought experiment was as impressive as it was simple.

The second thought experiment, which Galileo attributes to Aristotle's followers, is somewhat more thorough in drawing out the very same conclusion.

> SALVIATI: [Here is what] is considered an irrefutable argument for the earth being motionless. For if it made the diurnal rotation, a tower from whose top a rock was let fall, being carried by the whirling of the earth, would travel many hundreds of yards to the east in the time the rock would consume in its fall, and the rock ought to strike the earth that distance away from the base of the tower. This effect [the Aristotelians] support with another experiment, which is to drop a lead ball from the top of the mast of a boat at rest, noting the place where it hits, which is close to

the foot of the mast; but if the same ball is dropped from the same place when the boat is moving, it will strike at that distance from the foot of the mast which the boat will have run during the time of fall of the lead, and for no other reason than that the natural movement of the ball when set free is in a straight line toward the center of the earth.

(*Dialogo*, 126)

You say, then, that since when the ship stands still the rock falls to the foot of the mast, and when the ship is in motion it falls apart from there, then conversely, from the falling of the rock at the foot it is inferred that the ship stands still, and from its falling away it may be deduced that that ship is moving. And since what happens on the land, from the falling of the rock at the foot of the tower one necessarily infers the immobility of the terrestrial globe.

(*Dialogo*, 144)

In both thought experiments, the analogies between the moving boat and the 'moving earth' seem quite appropriate. Thus the only chance of refuting them lies in challenging the central assumption. After all, why should a projectile launched from a moving object 'fall behind'? According to Aristotelian dynamics, it is 'for no other reason than that the natural movement of the ball when set free is in a straight line toward the center of the earth' (*Dialogo*, 144). Oresme and Galileo disagree. Both challenge that assumption and propose an alternative framework—one in which a projectile launched from a moving object in some sense preserves the motion that was previously imparted to it.

Oresme directly undermines the central assumption of where the body would fall by positing that if the earth were rotating, it would carry the atmosphere along with it. (This anticipates 'aether drag' assumptions that would be made in the late nineteenth century in order to account for unexpected results in optical experiments.) Hence, if an arrow were to be shot vertically up, the moving air would be responsible for keeping the arrow on track as it falls back towards its point of origin. Thus, just from seeing the stone return to its place of origin, we can conclude nothing about the motion of the earth.

Concerning the arrow or stone thrown upwards, etc., one could say that the arrow fired up is moved very quickly towards the east with the air through which it passes, and with all the mass of the lower part of the world which is moved with a diurnal movement; and for this reason, the arrow falls back to the place on the earth from which it started.

(Oresme, *Le Livre du ciel et du monde*, in Sambursky, 162)

The mere fact that a projectile launched vertically upwards returns to its place of origin confirms nothing. It could be the result of the earth standing still, or it could just as easily be explained in a Copernican model with atmospheric drag. The point is that Ptolemy's thought experiment fails, because it does not take into account something like the principle of Galilean relativity. Oresme, however, clearly does possess at least an intuitive appreciation of relative motion, if not the full-fledged idea.

The real issue here is the relativity of motion, not antiperistasis. Even conceding, for the sake of the argument that Aristotle is right in his dynamics, the problem of rotation of the earth is still not settled. Galileo makes the same point. The Copernican model of the universe as such is not incompatible with the tenets of Aristotelian dynamics. All one has to do is posit a medium which is normally responsible for projectile motion that also rotates along with the earth. Hence, the moving air itself ensures that the stone 'follows along' with the tower's horizontal motion.

> SALVIATI: I might add at least that part of the air which is lower than the highest mountains must be swept along and carried around by the roughness of the earth's surface, or must naturally follow the diurnal motion because of being a mixture of various terrestrial vapors and exhalations. That [object] which leaves the top of the tower finds itself in a medium which has the same motion as the entire terrestrial globe, so that far from being impeded by the air, it rather follows the general course of the earth with assistance from the air.
>
> (*Dialogo*, 143)

In order to render this thesis more credible, Galileo's next thought experiment draws upon an analogy between a flying eagle and the moving earth. According to Galileo, if the wind imparts some horizontal motion to the eagle's dropped payload, it can similarly be responsible for the 'slanting' parabolic trajectory of a stone dropped from a tower. Notice that Galileo speaks of using the 'mind's eye' in order to envision this analogy—he is explicitly presenting this argument as a thought experiment.

SALVIATI: If you want to present a more suitable experiment, you ought to say what would be observed (if not with one's actual eyes, at least with those of the mind) if an eagle, carried by the force of the wind, were to drop a rock from its talons. Since this rock was already flying equally with the wind, and thereafter entered into a medium moving with the same velocity, I am pretty sure that it would not be seen to fall perpendicularly, but, following the course of the wind and adding to this that of its own weight, would move in a slanting path.

(*Dialogo*, 143)

It is important to point out that Galileo himself does not intend to rely too heavily on this last analogy. After all, as we shall see in a moment, he strongly opposes the view that the medium is responsible for projectile motion. Rather, Galileo merely wishes to point out that the trajectory is not always what it seems. Thus, even a committed Aristotelian must admit that, while from the point of view of the eagle, the dropped payload perhaps seems to fall in a straight vertical line downwards, an earthbound observer sees the rock trace out a parabolic path. If the Aristotelian wants, he can attribute this motion to the wind rushing behind the eagle.

Oresme's response to Ptolemy includes several short thought experiments which confirm his thesis by analogy.

And [the trajectory of the arrow] seems possible if one considers a similar case; suppose a man to be in a ship travelling very swiftly towards the east, without perceiving this motion, and suppose him to move his hand quickly down against the mast of the ship, it would seem to him that his hand was not moved except in a straight line; and thus, according to this opinion, the case of the arrow which rises or falls straight up or straight down seems to us similar.

(Oresme, *Le Livre du ciel et du monde*, Sambursky, 162–3)

Oresme identifies a fault in the central Ptolemaic and Aristotelian assumption: that only in a frame of reference 'absolutely' at rest could the laws of motion conform to our experience. The faulty premiss should be replaced, Oresme urges, with an understanding of motion as a purely relative phenomenon. Oresme thus leaves us in an agnostic state, unable to choose between the rival accounts—Ptolemaic or Copernican. He offers us a counter thought experiment that has the effect of neutralizing the first without resolving the issue in favour of the alternative. When it comes to the rotation of the earth, the fault is not so much with the dynamic theory (including antiperistasis), rather it is a failure to recognize the need for something akin to the relativity of motion.

Counter thought experiments have a rather complex structure. The experiential part is in the middle. Typically, there is a set-up, followed by a claim of some phenomenon, followed by an explanation, which we tend to think of as the result of the thought experiment. What counter thought experiments really do is undermine some other thought experiment that has previously been found persuasive. (See chapter three for details.) Thus, first came Newton's bucket thought experiment, which was used by him to argue for absolute space. We start with a bucket half-filled with water. It is connected to a twisted rope, released, and starts to spin. The surface of the water, initially flat and at rest with respect to the bucket, eventually becomes concave. What is causing this? Newton's answer is that the water-bucket system (every part of which is at rest with respect to every other part), must be moving—rotating—with respect to space itself. And, so, space must be a thing in its own right.

Mach later produced a counter thought experiment by asserting that the water in a bucket with extremely thick walls would *not* climb the walls, if the bucket were rotated. Of course, we have no reason to believe Mach's assertion, but we see from his counter thought experiment that Newton's claims about what would happen to the bucket in empty space are not so obvious and inevitable as we had previously thought. We need not accept Mach's positive conclusion that all motion must be relative motion, but we do find that his counter thought experiment has undermined Newton's. And that is precisely what Oresme's thought experiment has done to Ptolomy's.

Let's turn now to Galileo. We will see that although Galileo's thought experiments proceed much in the same vein as Oresme's, they provide a more rigorous justification for their assumptions. The first thought experiment does not introduce a new analogy, but, like Oresme's first argument, instead challenges Aristotle's underlying dynamical assumption (i.e., that if the earth were rotating, a projectile launched vertically up would 'trail behind' its point of origin). It, too, is a counter thought experiment.

> SALVIATI: But if it happened that the earth rotated, and consequently carried along the tower, and if the falling stone were seen to graze the side of the tower just the same, what would its motion have to be? . . . The motion would be a compound of two motions; the one with which it measures the tower, and the other with which it follows it. From this compounding it would follow that the rock would no longer describe that simple straight perpendicular line, but

a slanting one, and perhaps not straight . . . Hence just from seeing the falling stone graze the tower, you could not say for sure that it described a straight and perpendicular line, unless you first assumed the earth to stand still.

(*Dialogo*, 139–40)

Galileo clearly explains that the mere observation of a falling rock on its own is inconclusive. He is obviously endorsing the relativity of motion. Motion can only be perceived relative to one's own inertial reference frame. That is, no object moving at the same velocity as an observer can help that observer determine whether or not she is actually moving. No 'strange' kinematic effects occur within frames of reference moving with uniform velocity. To illustrate the relativity of motion, and, by analogy, the futility of using earth-bound experiments to determine whether or not the earth itself is rotating, Galileo presents the following thought experiment, quoted before. It is one of his most effective:

SALVIATI: Shut yourself up with some friend in the main cabin below decks on some large ship, and have with you there some flies, butterflies and other small flying animals. Have a large bowl of water with some fish in it; hang up a bottle that empties drop by drop into a wide vessel beneath it. With the ship standing still, observe carefully how the little animals fly with equal speed to all sides of the cabin. The fish swim indifferently in all directions; the drop falls into the vessel beneath; and, in throwing something to your friend, you need throw no more strongly in one direction than another, the distances being equal; jumping with your feet together, you pass equal spaces in every direction. When you have observed all these things carefully (though there is no doubt that when the ship is standing still everything must happen in this way), have the ship proceed with any speed you like, so long as the motion is uniform and not fluctuating this way and that. You will discover not the least change in all the effects named, nor could you tell from any of them whether the ship was moving or standing still.

(*Dialogo*, 186)

What a wonderful thought experiment it is. It generates that much sought after Aha! effect or Eureka moment when all becomes clear. This thought experiment, perhaps more than any other, yields the idea of the relativity of inertial frames of reference, linked with what came to be called Galilean relativity.

Armed with this principle of relativity, Galileo is now able to reverse the original analogy which the Aristotelian thought experiment drew upon in an attempt to establish the immobility of the earth.

SALVIATI: If the stone dropped from the top of the mast when the ship was sailing rapidly fell in exactly the same place on the ship to which it fell when the ship was standing still, what use could you make of this falling with regard to determining whether the vessel stood still or moved?

(*Dialogo*, 144)

To his own question, he again offers an agnostic response:

SALVIATI: The same cause holding good on the earth as on the ship, nothing can be inferred about the earth's motion or rest from the stone falling always perpendicularly to the foot of the tower.

(*Dialogo*, 145)

Impetus and inertia

The crucial change in the post-Aristotelian outlook is the transfer of the source of motion from the medium to the object itself. Galileo argues in this next thought experiment that the moving ship itself imparts a horizontal motion to the rock as it sits on top of the mast. As the rock falls, it preserves this horizontal motion (since there is nothing to impede it) and therefore lands straight below its point of departure.

SALVIATI: As to that stone which is on top of the mast; does it not move, carried by the ship . . .? And consequently is there not in it an ineradicable motion, all external impediments being removed? And is not this motion as fast as that of the ship?

SIMPLICIO: By the final conclusion you mean that the stone, moving with an indelibly impressed motion, is not going to leave the ship, but will follow it, and finally will fall at the same place where it fell when the ship remained motionless.

(*Dialogo*, 148)

What led up to this? Let us begin with a selection of remarks from Buridan, amounting to his theory of impetus. According to Buridan, impetus is the motive force which is responsible for maintaining detached motion, whether projectile or circular.

If light wood and heavy iron of the same volume and of the same shape are moved equally fast by a projector, the iron will be moved farther because there is impressed in it a more intense impetus, which is not so quickly corrupted as the lesser impetus would be corrupted.

This also is the reason why it is more difficult to bring to rest a large smith's mill which is moving swiftly than a small one, evidently because in the large one, other things being equal, there is more impetus.

And for this reason you could throw a stone of one-half of one pound weight farther than you could a thousandth part of it. For the impetus in that thousandth part is so small that it is overcome immediately by the resisting air.

(Buridan, *Questions on the Physics of Aristotle* VIII.12.5,
in Clagett 1961, 535)

We should stress the very thing that Buridan stresses, namely, the source of motion is in the object, not the surrounding medium. In fact, the medium offers some resistance, which depletes the impetus. Buridan goes so far as to say what would happen if all sources of resistance were to be removed from a spinning wheel.

If you cause a large and very heavy smith's mill [i.e., a wheel] to rotate and you then cease to move it, it will still move a while longer by this impetus it has acquired. Nay, you cannot immediately bring it to rest, but on account of the resistance from the gravity of the mill, the impetus would be continually diminished until the mill would cease to move. And if the mill would last forever without some diminution or alteration of it, and there were no resistance corrupting the impetus, perhaps the mill would be moved perpetually by that impetus.

(1961, 561)

Now Buridan can definitively conclude that Aristotelian dynamics are false. For recall that in Aristotle's theory, motion without resistance is impossible, because then an object would move with infinite velocity. Buridan thus presents a logically plausible scenario which, he gives us reason to believe, would result in motion of eternal duration—indeed inconsistent with Aristotelian dynamics.

Buridan develops a similar thought experiment in order to explain the motion of celestial bodies, which subsist in an utterly resistance-free environment. Again, the thought experiment consists of an inference to the best explanation: the reason why the motion of celestial bodies appears perpetual is because they encounter no resistance to 'corrupt' the impetus originally imparted on them.

You know that many maintain that a projectile, after leaving the thrower, is moved by impetus imparted by the thrower and moves as long as the impetus remains stronger than the resistance; and that the impetus would last forever if it were not diminished and destroyed by the opposing resistance or by the tendency to contrary motion. And in celestial motions there is no opposing resistance; therefore when God, at the Creation, moved each sphere of the heavens with just the velocity he wished, he [then] ceased to move them himself, and since then those motions have lasted forever due to the impetus impressed on those spheres.

(Buridan, *Questions on the Metaphysics of Aristotle* XII.9, in Maier 1982a, 89)

In a particularly Ockham's razor-like manoeuvre, Buridan argues that the celestial bodies in and of themselves contain the reason for their continued motion. There is no need to posit additional causes or 'intelligences' in order to explain cosmological movement.

It does not appear necessary to posit intelligences of this kind, because it could be answered that God, when He created the world, moved each of the celestial orbs as He pleased, and in moving them impressed in them impetuses which moved them without His having to move them any more . . . And these impetuses which He impressed in the celestial bodies were not decreased or corrupted afterwards because there was not inclination of the celestial bodies for other movements. Nor was there resistance which would be corruptive or repressive of that impetus.

(Buridan, *Questions on the Physics of Aristotle* VIII.12, in Clagett 1961, 524–5)

Let's turn to Oresme and a pair of thought experiments in the same spirit as Buridan's, in support of the theory of impetus. A wheel that is set in motion

cannot be stopped immediately without great difficulty, but is moved by a certain impetus, though separated from the first moving agent and [is not moved] by the surrounding air.

(Oresme, *Questiones super De celo* II.7, in Grant 2002, 155)

Having progressively eliminated all possible sources of motion, Oresme concludes that the spinning wheel must have received an impetus which effectively 'keeps it going'. For this is the best remaining possible explanation.

Oresme's second thought experiment is more complex, but also more successful because it sets up an idealized scenario in a resistance-free environment. It thus enables Oresme to scrutinize more thoroughly the consequences of the theory of impetus. Much like his predecessor, Buridan, Oresme infers that in the absence of resistance, the impetus imparted on an object will keep it moving forever. This passage appears in Oresme's *Le Livre du ciel et du monde*, where he discusses incommensurable circular motion.

> Now, I should like to show next that it is possible both in fact and in uncontradictable theory that some motion has a beginning and lasts forever. First, with regard to circular motion: I assume that a wheel of any kind of material is like the wheel of a clock [see Oresme's figure]; this we will designate as *a* and will call its center *a*. Let us next set in *a* another wheel smaller than the first, in the manner of an epicycle, with its center labeled *b*. Let us add a third wheel set on the second, having its center on the circumference of the second like the moon in its epicycle, and let us call this third wheel *c*. Now let us place a fourth wheel outside these three so fixed that *c* can touch it, and let its center be marked *d*. Then, I posit that *a* be moved around its center, that *b* be moved with the motion of *a* on which it is set, and, with this, that *b* be moved also with its proper motion around its center, while *c* has no proper motion but is moved with the motion of *a* and *b* . . .

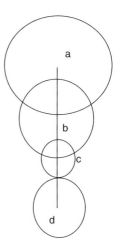

> Now I posit that *d* be so adjusted or controlled by counterweights and otherwise that it is inclined to move but is not moved until touched by *c*; this contact removes the hindrance and *d* begins to move regularly. Something of this sort or similar to it could be performed artificially or by skilled craftsmanship. Next, I posit that the two motions of *a* and *b* should be incommensurable, regular, and perpetual . . . At this present moment, I posit that the wheel *c* touches wheel *d*, and I say that it is impossible that it should have touched it previously or should touch it again, for it can touch it only when the centers *b*, *c*, and *d* are in conjunction exactly in one line; and this cannot have happened before not again in future time . . . From this it follows necessarily that wheel *d* would now begin to move and would never stop; and, although such a series of

events cannot occur in nature, nor be shown by material art or in destructible matter, nor endure so long [as forever], nevertheless, it contains or implies no contradiction whatever, nor is it within its own frame of reference incongruous to reason, but it is possible if we grant the nature of the motions. All the incongruity arises by reason of the material or from something outside the frame of reference.

(Oresme, quoted in Grant 2002, 38)

Perhaps the most effective example of an inference to the best explanation occurs in a thought experiment Galileo uses to illustrate the law of inertia. Because the example is so abstract, Galileo is able to minimize and simplify the assumptions needed in order to infer the conclusion.

SALVIATI: Suppose you have a plane surface as smooth as a mirror and made of some hard material like steel. This is not parallel to the horizon, but somewhat inclined, and upon it you have placed a ball which is perfectly spherical and of some hard and heavy material like bronze. What do you believe this will do when released?

SIMPLICIO: [It rolls down.]

SALVIATI: Now how long would the ball continue to roll, and how fast?

SIMPLICIO: The ball would continue to move indefinitely, as far as the slope of the surface extended, and with a continually accelerated motion.

SALVIATI: If one wanted the ball to move upward on this same surface, do you think it would go?

SIMPLICIO: Not spontaneously, no; but drawn or thrown forcibly, it would.

SALVIATI: Now tell me what would happen to the same movable body placed upon a surface with no slope upward or downward.

SIMPLICIO: There being no downward slope, there can be no natural tendency toward motion; and there being no upward slope, there can no resistance to being moved, so there would be an indifference between the propensity and the resistance to motion. Therefore it seems to me that it ought naturally to remain stable.

SALVIATI: But what would happen if it were given an impetus in any direction?

SIMPLICIO: It must follow that it would move in that direction.

SALVIATI: How far would you have the ball continue to move?

SIMPLICIO: As far as the extension of the surface continued without rising or falling.

SALVIATI: Then if such a space were unbounded, the motion on it would likewise be boundless? That is, perpetual?

SIMPLICIO: It seems so to me, if the movable body were of durable material.

(*Dialogo*, 145–7)

In this Socratic-like dialogue, Salviati draws out of Simplicio the conclusion that, in the absence of resistance or indeed any externally-acting force, there is no need for an object to change its motion. In his thought experiment, Galileo uses a step-by-step method to progressively eliminate all the influences which the environment might realistically have on the ball. First of all, he asks us to imagine the ball and plane surface as perfectly smooth and without any impediment—so that all sources of friction might be removed from the thought experiment. Next, Galileo has the surface held at such an angle so as to cancel any effect which gravity might have on the ball. Finally, we are told that the surface is unbounded and extends to an infinite distance in every direction. By abstracting from all resistances and dynamical influences on the ball, Galileo concludes that the speed of the ball will forever remain unchanged. This result is in clear opposition to Aristotelian dynamics, but thanks to Galileo's clever manipulation of the dynamical variables on the ball, we cannot help but infer that it must be true. Notice that Galileo's dynamical explanation of the phenomenon of inertial motion is just to say that the ball has neither a propensity nor a resistance to motion. Thus, Galileo turns the tables on the question which preoccupied both Aristotle and his mediaeval critics. Instead of assuming that the ball's natural state is one of rest, and asking 'what *keeps* the ball in uniform motion', he assumes that the ball's natural state is one of uniform motion, and asks 'what *prevents* the ball from staying in uniform motion'.

The inference to the best explanation seems in general to be an effective style of thought experiment. Though it all comes as a package, strictly, the thought experiment establishes some phenomenon, then after that comes the explanation of it. The effectiveness of this form of reasoning is perhaps unsurprising, given how closely this kind of thought experiment resembles the style of scientific reasoning used in actual empirical experiments.

Conclusion

The story of inertial motion does not end here, but what remains must be left to other occasions. For now a brief look ahead will have to suffice, starting with another example from Galileo.

> I mentally conceive of some moveable projected on a horizontal plane, all impediments being put aside. Now it is evident from what has been said elsewhere at greater length that equable motion on this plane would be perpetual if the plane were of infinite extent.
>
> (*Discorsi*, 170)

By infinite extent, Galileo can mean no more than unbounded, that is, there is no impediment to the object's continued motion. He does not intend an infinite straight line, since Galileo's inertia is circular. Centres of gravity are all important in his conception of how and why things move. Imagining a body in horizontal motion, he declares,

> . . . it is impossible that a heavy body . . . should naturally move upward, departing from the common center toward which all heavy bodies mutually converge; and hence it is impossible that these be moved spontaneously except with that motion by which their own center of gravity approaches the said common center. Whence, on the horizontal, which here means a surface [everywhere] equidistant from the said [common] center, and therefore quite devoid of tilt, the impetus or momentum of the moveable will be null.
>
> (*Discorsi*, 172)

Descartes deserves the credit for the modern concept of inertia, namely, a state of motion that involves constant speed in a straight line.

> The first law of nature: that each thing, as far as it is in its power, always remains in the same state; and that consequently, when it is once moved, it always continues to move.
>
> (Descartes, *Principles of Philosophy*, II, 37)

> The second law of nature: that all movement is, of itself, along straight lines; and consequently, bodies which are moving in a circle always tend to move away from the centre of the circle which they are describing.
>
> (*Principles of Philosophy*, II, 39)

Descartes, in these passages, has introduced two crucially important things. First, the idea of a *state*. Even though a body may be changing its location, it's state could be constant. Second, the idea that inertial motion is along straight lines.

By the time we get to Newton, inertial motion is well entrenched and shows up as the first law in Newton's *Principia*. The remarkable thing, from our point of view, is that by the time Newton formulated his view, he was putting inertia to work in his thought experiments for absolute space. Why does the water climb the sides of the bucket? The parts of water are rotating, that is, they are accelerating with respect to space itself. Space is the cause of inertial motion of a body, that is, of constant speed in a straight line, when there is no force being applied. With Newton, we have come back to the idea that some aspects of motion being due to the medium, not the body. It is space itself that is the cause of constant velocities when no forces are acting.

After Newton, the next big event is the re-affirmation of a principle of relativity in the special theory of relativity. It had been seriously threatened by nineteenth-century electrodynamics, where it was thought that the aether frame is the one and only frame in which the laws of nature hold. Following that, the general theory of relativity introduced free fall, that is, motion along space-time geodesics as inertial motion, the natural state for any body. It is impossible to say what the future holds for these notions. When a theory of quantum gravity finally emerges, we may find that space and time themselves have disappeared and with them the very idea of motion.

7 Einstein's brand of verificationism

No one better exemplifies the magic, mystery, and awesome might of physics than does Albert Einstein. The unruly hair, the baggy pants, the Germanic accent, these in the public mind (or at least in my mind when growing up) are the characteristics of genius. For philosophers, too, he is a hero, for he seems distinctly like one of us when he declares that 'Science without epistemology is—in so far as it is thinkable at all—primitive and muddled' (1949, 684).

But the source of the appeal goes beyond this. Einstein is something of a man for all seasons; we can find him catering to every philosophical taste. Those who like their physics *a priori* are delighted to find Einstein the old-fashioned rationalist who holds 'pure thought can grasp reality' (1933, 274). But staunch empiricists can take heart, too, since he can also be found saying 'Pure logical thinking cannot yield us any knowledge of the empirical world; all knowledge of reality starts from experience and ends in it' (1933, 271).

Einstein was a great thought experimenter; only Galileo was his equal. We have already seen a number of his creations in previous chapters. Now I want to examine their role in his physics, especially in relation to his general views on the epistemology of science. This really means we have two tasks before us; the harder is to say just what his general epistemological outlook was; once that is settled, the second task of fitting in thought experiments is relatively easy.

Fall of the positivist image

For a very long time an empiricist picture of Einstein has been dominant. The odd remark by the older Einstein looking back on his early work has contributed, but the main reason for this view has been the way the theories of special and general relativity were initially presented—both smacked of verificationism. And if Einstein did not make explicitly detailed philosophical pronouncements along positivist lines, well, even that did not really matter too much to empiricist commentators, since, as Reichenbach put it, 'It is not necessary for him to elaborate on it . . . he merely had to join a trend . . . and carry [it] through to its ultimate consequences' (1949, 290).

But this positivist picture of Einstein has largely fallen by the wayside in the last few years. Now it is a commonplace to view him as an empiricist in his early days who became a realist in his maturity. No one has done more to create this new and highly attractive picture than Gerald Holton, who describes Einstein's philosophical development as 'a pilgrimage from a philosophy of science in which sensationalism and empiricism were at the center, to one in which the basis was a rational realism' (1968, 219). And Einstein himself is quite obliging; in his 'Auto-biographical Notes' he seems to paint the same developmental picture. There he remarks that Mach undermined his early naivety, but that he adopted the great Austrian philosopher-physicist's brand of positivism *only* in his youth; eventually, he tells us, he came to see its shortcomings and dropped it.

> It was Ernst Mach who, in his *History of Mechanics,* shook this dogmatic faith; this book exercised a profound influence upon me in this regard while I was a student. I see Mach's greatness in his incorruptible scepticism and independence; in my younger years, however, Mach's epistemological position also influenced me very greatly, a position which today appears to me to be essentially untenable. For he did not place in the correct light the essentially constructive and speculative nature of thought.
>
> (1949, 21)

Others besides Holton attribute the developmental view to Einstein. Arthur Miller in his recent studies (1981, 1984) and Arthur Fine in *The Shaky Game* (1986) are two prime examples. Fine provides an interesting contrast with Holton. Both see Einstein 'turning away from his positivist youth . . .', as Fine puts it, 'and becoming deeply committed to realism' (1986, 123); but Holton sees this as a definite move in the right direction while Fine, on the other hand, tends to downplay Einstein's later realism and instead glories in his youthful empiricism. 'Einstein's early positivism and his methodological debt to Mach (and Hume) leap right out of the pages of the 1905 paper on special relativity. The same positivist strain is evident in the 1916 general relativity paper as well . . . .' And Fine leaves no doubt that he takes this anti-realism to be a great virtue: '. . . it would be hard to deny the importance of this instrumentalist/positivist attitude in liberating Einstein from various realist commitments. Indeed', Fine continues, 'without the "freedom from reality" provided by his early reverence for Mach, a central tumbler necessary to unlock the secret of special relativity would never have fallen into place' (1986, 122f.).

The developmental view of Einstein is enormously attractive. It seems to do justice to Einstein's own autobiographical remarks, and even better, it fits in nicely with the temper of contemporary times. Let's face it, positivism is dead and, in spite of recalcitrants like Fine and van Fraassen, most of us are realists.[1] Isn't it nice that the greatest scientist of the century is one of us? Oh yes, we might add, Einstein was a positivist in his early days, but he soon saw through that and became a scientific realist.

It is hard not to be attracted to this developmental picture, but there are difficulties.

Problems with the developmental picture

For all its appeal, the developmental account runs into difficulties on several fronts. Here are some of the problems with thinking Einstein made a 'pilgrimage' from positivism to realism:

(1) One of the most convincing considerations for thinking Einstein was a positivist in his youth is the formulation of special relativity, with all its talk about rods and clocks, etc. But we must not forget that special relativity has lots of non-observable features; for instance, it postulates an infinite class of inertial frames, something very far from experience.

(2) During the same *annus mirabilis* that he produced his paper on special relativity, Einstein also published two other great works on Brownian motion (1905b) and light quanta (1905c). Later Einstein rightly said of his Brownian motion paper that it 'convinced the sceptics ... of the reality of atoms' (1949, 49). This is hardly the scientific work of a true Machian positivist—yet, it was produced at the same time as special relativity.

(3) The decline of empiricism has not had a detrimental effect on either special or general relativity. If these theories are indeed linked to extreme empiricism as, say, phenomenological thermodynamics or psychological behaviourism are linked to empiricism, then we might expect relativity to have justly fallen on hard times—but it hasn't. So any connection between relativity and positivism is superficial at best.

(4) Developmentalists offer little or nothing in the way of an explanation for Einstein's alleged philosophical change of heart.

Holton suggests that Einstein's later realism came about with a growing religiosity; he refers to the 'connections that existed between Einstein's scientific rationalism and his religious beliefs'. And Holton further remarks that There is a close tie between his [Einstein's] epistemology, in which reality does not need to be validated by the individual's sensorium, and what he [Einstein] called "Cosmic Religion'" (Holton 1968, 242f.). But this is quite unhelpful, since Einstein was never very serious about religious matters—he tended to use religious metaphors, such as 'God does not play dice', the way atheists use 'God's-eye point of view'; and anyway, to say Einstein was becoming spiritual is really nothing better than a slightly mystified way of saying he became a realist. It certainly explains nothing. Duhem, by contrast, was an anti-realist because of his religious beliefs.

(5) Einstein's alleged new-found realism is used to explain his objection to quantum mechanics. However, this attempted explanation runs together two different senses of realism which I will explain below.

(6) Holton sometimes makes Einstein out to be a nonverificationist even in 1905 when constructing his special relativity paper. He says of Einstein's principle of relativity that it was 'a great leap . . . far beyond the level of the phenomena' (1981, 89). Of course, it is possible to follow Elie Zahar on this when he remarks that 'while paying lip service to Machian positivism, scientists like Einstein remained old-fashioned realists' (1977, 195). But then the developmental view is trivialized; the only change in Einstein is that by becoming an explicit realist he came to hold a more accurate view of what he had been doing all along. But the cost of such an interpretation is considerable: we lose the explanatory power to account for various features present in much of Einstein's early scientific work, empiricist-like features which are definitely there.

(7) Related to this is Holton's explanation of why Mach, much to Einstein's surprise, denounced relativity. Holton thinks it was because Mach saw through it and realized just how realistic and anti-empiricist the theory of relativity actually was. However, thanks to the recent detective work of Gereon Wolters (1984), we now know that Mach's 'rejection' was actually the forgery of his son Ludwig Mach. So we no longer need to explain away Mach's antipathy; indeed, just the opposite.

(9) Most importantly perhaps, there is no mention in the developmental account of Einstein's distinction between 'principle' and 'constructive' theories, a distinction which he seems to have thought quite important. It turns out that the theory of relativity is a principle theory while quantum mechanics is a constructive one. The illusion

of a philosophical change from positivist to realist is fostered, I will suggest, by the fact that Einstein's philosophical remarks focused on relativity during the early part of his career, while his philosophical attention changed to quantum mechanics in his maturity; this was not a change in philosophical outlook so much as a change in the subject of his interest.

(10) And finally, what about thought experiments? Consistently throughout his long career Einstein brilliantly employed thought experiments. The developmental picture of Einstein leaves this central ingredient in his *style* of thinking completely unaccounted for. Of course, the pilgrimage view might still be right without having anything in particular to say about thought experiments, but a different account which fits them in must surely be preferred.

In light of these many problems with the developmental account, a quite different picture of Einstein's philosophical views seems called for. I'll now try to construct a plausible view which, among other things, tries to do some justice to thought experiments.

Principle and constructive theories

Einstein liked to distinguish between two types of theories in physics, *principle theories* and *constructive theories*. The latter type of theory is any kind of hypothesis or conjecture which is put forward to explain a wide variety of facts.

> [Constructive theories] attempt to build up a picture of the more complex phenomena out of the materials of a relatively simple formal scheme from which they start out. Thus the kinetic theory of gases seeks to reduce mechanical, thermal, and diffusional processes to movements of molecules—i.e., to build them up out of the hypothesis of molecular motion. When we say that we have succeeded in understanding a group of natural processes, we invariably mean that a constructive theory has been found which covers the processes in question.
>
> (Einstein 1919, 228)

A principle theory, on the other hand, starts with something known to be true (for example, the speed of light in a vacuum is constant) and then forces everything else to conform to this principle. Unlike constructive theories which are speculative, explanatory, and attempt to unify diverse phenomena, principle theories never try to explain anything.

The elements which form their basis and starting-point are not hypothetically constructed but empirically discovered ones, general characteristics of natural processes, principles which give rise to mathematically formulated criteria which the separate processes or the theoretical representations of them have to satisfy. Thus the science of thermodynamics seeks by analytical means to deduce necessary conditions, which separate events have to satisfy, from the universally experienced fact that perpetual motion is impossible.

(1919, 228)

Einstein goes on to contrast these two types of theory and tells us which type relativity is.

The advantages of the constructive theory are completeness, adaptability, and clearness, those of the principle theory are logical perfection and security of the foundation.

The theory of relativity belongs to the latter class.

(1919, 228)

I am going to use Einstein's distinction between principle and constructive theories to paint a different picture of his philosophical views than the one given by the developmental account. Einstein's verificationism, I suggest, applies only to his principle theories, not to his constructive ones, where he was arguably some sort of realist. Thus, it is no surprise to see positivist-sounding language in special and general relativity, for example, but not in his work on Brownian motion or light quanta. On the other hand, Einstein *appears* to have dropped his early empiricism and become a realist. I shall maintain, however, that this sort of change in philosophical view did not really occur—indeed, there was very little change at all. Rather, what did happen was a change in focus; his early attention was on relativity, a principle theory, while later it was on quantum mechanics, a constructive theory. There was a change in his scientific interests, but Einstein's philosophical views remained fairly stable throughout his life.

I must add, however, that the distinction between principle and constructive theories is not a sharp one. If it is a useful distinction—and I think it is—it must be understood as somewhat fuzzier than Einstein might have desired. The distinction is perhaps best understood by analogy with the more familiar distinction between the observable and the theoretical. This latter distinction is not a sharp one either, but clear examples on either side of the boundary exist. Trees, rabbits, unicorns, and pointer readings are observable, while electrons, genes, phlogiston, and superegos are theoretical. (Notice that unicorns are 'observable', but not 'observed', which is why we think there are none.)

Einstein characterizes principle theories as 'secure' and as 'non-explanatory', while constructive theories to the contrary are both explanatory and highly conjectural, hence insecure. By rejecting a sharp distinction between principle and constructive theories, we in effect reject a sharp distinction between explanatory and non-explanatory theories, between conjectural and non-conjectural theories, and between secure and insecure theories. These considerations will come up again below.

Free creations of the mind

Einstein is famous, or as some would have it, infamous, for his resistance to the quantum theory. There are two responses people typically have made to his resistance (and to some extent there is a tension between those two responses). One is to dismiss Einstein as an old dog who couldn't learn new tricks. The other response is to express puzzlement at Einstein's resistance, since it was thought that the quantum theory, after all, was just a natural result of that same philosophical attitude that Einstein himself applied so successfully in the founding of relativity (i.e., not really a new trick after all). This latter view is nicely illustrated in the exchange between Einstein and Heisenberg as recounted by Heisenberg himself:

> 'But you don't seriously believe,' Einstein protested, 'that none but observable magnitudes must go into a physical theory?'
>
> 'Isn't that precisely what you have done with relativity?' I asked in some surprise. 'After all, you did stress the fact that it is impermissible to speak of absolute time, simply because absolute time cannot be observed; that only clock readings, be it in the moving reference system or the system at rest, are relevant to the determination of time.'
>
> Possibly I did use this kind of reasoning,' Einstein admitted, 'but it is nonsense all the same. Perhaps I could put it more diplomatically by saying that it may be heuristically useful to keep in mind what one has actually observed. But on principle, it is quite wrong to try founding a theory on observable magnitudes alone. In reality the very opposite happens. It is the theory which decides what we can observe . . .'

(1971, 63)

The philosophical position which so startled Heisenberg was a theme Einstein returned to and stressed again and again over the years.[2] Perhaps the first time it appears is in his address celebrating Planck's sixtieth birthday in 1918. Einstein's remarks are worth quoting at length as he not only outlines a conjectural or hypothetico-deductivist way of doing science, but also tackles the inevitable underdetermination problem:

> The supreme task of the physicist is to arrive at those universal elementary laws from which the cosmos can be built up by pure deduction. There is no logical path to these laws; only intuition, resting on sympathetic understanding of experience, can reach them. In this methodological uncertainty, one might suppose that there were any number of possible systems of theoretical physics all equally well justified; and this opinion is no doubt correct, theoretically. But the development of physics has shown that at any given moment, out of all conceivable constructions, a single one has always proved itself decidedly superior to all the rest. Nobody who has really gone deeply into the matter will deny that in practice the world of phenomena uniquely determines the theoretical system, in spite of the fact that there is no logical bridge between phenomena and their theoretical principles . . .
>
> (1918, 226)

It is interesting to see Einstein coping with the underdetermination problem, and we must admire his optimism, if not his credulity.

In his most philosophically sustained work, 'Physics and Reality', which was written in the mid-1930s, Einstein outlines his view as follows:

> Physics constitutes a logical system of thought which is in a state of evolution, whose basis cannot be distilled, as it were, from experience by an inductive method, but can only be arrived at by free invention. The justification (truth content) of the system rests in the verification of the derived propositions by sense experiences . . .
>
> (1935, 322)

The style of reasoning that Einstein favours here, namely some sort of hypothetico-deductivism (H-D),[3] is one he was already employing in 1905. In the same 'Autobiographical Notes' in which he suggests he was a Machian, he also describes (correctly describes, I might add) his work in statistical mechanics. 'My major aim in this was to find facts which would guarantee as much as possible the existence of atoms of finite size' (1949, 47). Einstein goes on to outline the argument.

> The simplest derivation [of what turned out to be Brownian motion] rested upon the following consideration. If the molecular-kinetic theory is essentially correct, a suspension of visible particles must possess the same kind of osmotic pressure fulfilling the laws of gases as a solution of molecules. This osmotic pressure depends upon the actual magnitude of the molecules, i.e., upon the number of molecules in a gram-equivalent. If the density of the suspension is inhomogeneous, the osmotic pressure is inhomogeneous, too, and gives rise to a compensating diffusion, which can be calculated from the well known mobility of the particles.

He then concludes,

> The agreement of these considerations with experience ... convinced the sceptics . . . of the reality of atoms.
>
> (1949, 47f.)

There are several things to note in this passage. For one thing, Einstein describes himself as doing something quite anti-Machian in 1905, and so, *contra* Holton and Fine, Einstein is clearly a realist about theoretical entities in his youth.

The second thing to note is that the atomic theory is a constructive theory, and that Einstein's reasoning is clearly H-D. Many of Einstein's remarks, both to Heisenberg and in his various essays written in later life, seem to be grist to Holton's mill. Einstein's final philosophical position, including the rejection of verificationism and the adoption of some sort of realism, appears far from anything Mach would approve of. But such a conclusion is far too hasty. If we look back at Einstein's characterization of constructive theories we can see that he is simply calling for some sort of hypothetico-deductivism. This is especially clear in a letter Einstein wrote to his old friend Maurice Solovine (on 7 May 1952) in which he clarified his views with the help of a diagram (Solovine 1987, 137).

Much is made of this scheme by both Holton (1979) and Miller (1984) who quite rightly note that Einstein thinks there is a great gap between experience and the axioms. But we should be careful about the circumstances under which Einstein thinks a jump should be made. On Einstein's distinction, it is made in constructive, not in principle theories. Miller fatally misunderstands this distinction when he says

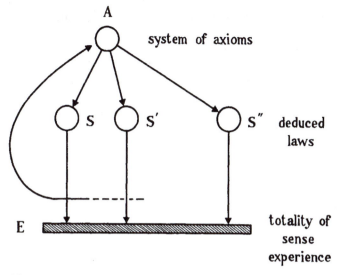

Figure 23

Einstein 'leaped across the abyss between these (E) to invent (A), which comprises the two principles of special relativity' (1984, 46). That is wrong. I want to pursue this, but first, a word about 'truth'.

Einstein's realism

For my purposes in this chapter it is not very important whether Einstein was a realist. Einstein's remarks cited above show him to be a hypothetico-deductivist. Holton, largely without argument, assimilates this with realism; but, of course, there is a considerable difference. One could hold with Duhem (1954) or van Fraassen (1980), for instance, that hypotheses are an essential part of science, but are merely useful fictions. Theories may employ any concept whatsoever; the only constraint is that they should 'save the phenomena'. Such an instrumentalism is certainly not realism, yet it is still a far cry from a Machian positivism which barred all but observable entities. Not all anti-realists are alike.

I am presuming that Einstein is a realist about his constructive theories, since I take it that the concepts which he liked to call 'free creations of the mind' are intended to refer and that the theories so constructed are actually true or false. However, this may not be correct. 'It is difficult to attach a precise meaning to "scientific truth",' says Einstein. 'Thus, the word "truth" varies according to whether we deal with a fact of experience, a mathematical proposition, or a scientific theory' (1929, 261).

Einstein does not elaborate here sufficiently, but we could imagine some form of anti-realism creeping in: observable 'facts' are 'true' in some ordinary correspondence sense, while theories are 'true' only in the instrumental sense that they are empirically adequate (i.e., imply all and only true facts). This seems also to be of a piece with remarks made many years later as part of his 'epistemological credo'.

> A proposition is correct if, within a logical system, it is deduced according to the accepted logical rules. A system has truth-content according to the certainty and completeness of its co-ordination possibility to the totality of experience. A correct proposition borrows its 'truth' from the truth-content of the system to which it belongs.
>
> (1949, 13)

Arthur Fine (1986) has with some justice recently challenged the view that Einstein is a regular scientific realist. For my purposes here it does not matter very much one way or the other. The real issue is this. Did Einstein develop from some sort of verificationism (in which there was a strict adherence to observable elements) to some sort of liberal H-D account (in which speculation and conjecture play a crucial role), and moreover, what was the character of his verificationism?

Relativity as a principle theory

Einstein repeatedly called relativity a principle theory. The starting point for such a theory, as he put it, is 'not hypothetically constructed but empirically discovered', and consequently has the 'advantage' of 'logical perfection and security of foundations' (1919, 228). Throughout his life, Einstein characterized both special and general relativity as non-speculative, non-hypothetical, non-conjectural, in short, as principle theories rather than constructive ones.

Writing to his friend Conrad Habicht in 1905 and sending him the fruits of his labours of that marvellous year, Einstein called his light quanta paper 'very revolutionary', while he merely noted that the relativity paper might be interesting in its kinematical part. Years later, *after* he had quite explicitly embraced an H-D view, Einstein was still claiming that relativity had a kind of verificationist justification.

I am anxious to draw attention to the fact that this theory [i.e., relativity] is not speculative in origin; it owes its invention entirely to the desire to make physical theory fit observed fact as well as possible. We have here no revolutionary act but the natural continuation of a line that can be traced through centuries.

(1921b, 246)

I want to contrast this with Einstein's H-D view of some other theories, a view which is normally identified as his mature view, but which he actually put forward in 1919, two years *before* the verificationist-sounding passage just cited.

The supreme task of the physicist is to arrive at those universal elementary laws from which the cosmos can be built up by pure deduction. There is no logical path to those laws . . .

(1919, 226)

It is clear from this pair of passages that Einstein is not making a pilgrimage from a Machian outlook to an H-D view—rather, he is holding both views simultaneously. How is this apparent contradiction possible? Simple. The H-D account applies to constructive theories and the Machian-sounding sentiments apply to principle theories such as relativity.

Holton makes the same sort of mistake that Miller makes when he calls the thinking that went into special relativity 'a conjecture', and

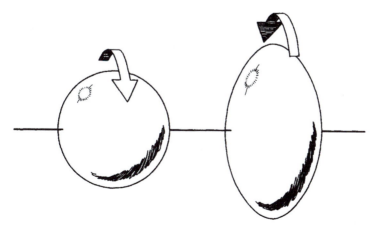

Figure 24

when he thinks of a constructive theory as 'one built up inductively from phenomena . . .' (1981, 88). Holton cites Einstein's remarks about the 'prinriple of relativity being raised to the status of a postulate' and calls it 'a great leap . . . far beyond the level of the phenomena . . .' (1981, 89). This shows a serious misunderstanding: first, of what a constructive theory is; second, of what the difference between constructive and principle theories is; and third, of what sort of theory relativity is. Let us turn now to the details involved in relativity to clear up some of these confusions.

Einstein's brand of verificationism

Einstein's positivism seems to 'leap right out of the pages', as Fine put it. He begins the general relativity paper (1916) with the remark that in classical mechanics there is an 'epistemological defect . . . pointed out by Ernst Mach' (1916, 112). Einstein then describes a thought experiment with two globes which are in observable rotation with respect to one another. One is a sphere, the other an ellipsoid of revolution. Einstein asks 'What is the reason for the difference in the two bodies?' He then sets verificationist conditions on any acceptable answer. I will quote at length, since the verificationism leads directly to Mach's principle and the principle of general co-variance.

> No answer can be admitted as epistemologically satisfactory, unless the reason given is an *observable fact of experience*. The law of causality has not the significance of a statement as to the world of experience, except when *observable facts* ultimately appear as causes and effects.

Einstein then declares that classical physics is not up to proper epistemological standards.

> Newtonian mechanics does not give a satisfactory answer to this question. It pronounces as follows:—The laws of mechanics apply to the space R1, in respect to which the body S1 is at rest, but not to the space R2, in respect to which the body S2 is at rest. But the privileged space R1 of Galileo, thus introduced, is a merely *factitious* cause, and not a thing that can be observed. It is therefore clear that Newton's mechanics does not really satisfy the requirement of causality in the case under consideration, but only apparently does so, since it makes the factitious cause R1 responsible for the observable difference in the bodies S1 and S2.

Einstein then goes on to say how things should be properly viewed, introducing both Mach's principle and the principle of general co-variance.

> The only satisfactory answer must be that the physical system consisting of S1 and S2 reveals within itself no imaginable cause to which the differing behaviour of S1 and S2 can be referred. The cause must therefore lie *outside* this system. We have to take it that the general laws of motion, which in particular determine the shapes of S1 and S2, must be such that the mechanical behaviour of S1 and S2 is partly conditioned, in quite essential respects, by distant masses which we have not included in the system under consideration. These distant masses and their motions relative to S1 and S2 must then be regarded as the seat of the causes (which must be susceptible to observation) of the different behaviour of our two bodies S1 and S2. They take over the role of the factitious cause R1. Of all imaginable spaces R1, R2, etc., in any kind of motion relatively to one another, there is none which we may look upon as privileged *a priori* without reviving the above-mentioned epistemological objection. The laws of physics must be of such a nature that they apply to systems of reference in any kind of motion.
>
> (1916, 112f. Einstein's italics throughout)

It is hard to resist the feeling that not only is this a strict form of empiricism, but that it is also, as Fine stresses, doing a great deal of valuable work. Einstein may well have been in some regards an 'old-fashioned realist', as Zahar (1977, 195) says, but it is most doubtful that he is here *merely* 'paying lip-service to Machian positivism'. There is a genuine Machian spirit to what is going on in both special and general relativity, and neither Einstein nor his commentators such as Holton and Fine are completely off target in describing it thus.

Of course, it seems absurd to see Einstein as both a realist and a verificationist simultaneously, but the tension is resolved when we see just what kind of verificationist he is. Einstein's brand of verificationism is not like any other; it gives special status to the intuitively obvious. His positivism is more an impulse to assign the self-evident a special role than to eliminate unobservable entities.

The identification of gravitational and inertial mass is a case in point. Here we have a type of unification based upon reflective considerations. It differs from what normally passes for unification: a theory which explains quite disparate phenomena is said to unify them. Such unification is also taken to be evidence that the unifying theory is true. However, the standard sort of unification, if it happens at all, is what goes on in Einstein's constructive theories. On the other hand, the unification which goes on in a principle theory like relativity is obviously different. It is a perceived, not a derived, unification; so it has no evidential merits. But then this is no surprise, since on Einstein's view, a principle theory doesn't explain anything anyway.

Principle theories are not *intended* to be explanatory; but, of course, we know they are—special relativity explains a lot. Let us try an analogy to help ease the apparent tension. From a straightforward empirical observation I am prepared to assert: 'The letter is under the cup on the table.' This assertion is not intended by me to be explanatory, but it does have explanatory consequences anyway. For example, 'Why didn't the wind blow the letter away?' My assertion explains why. The situation in this everyday case and in special relativity are similar. Since every proposition has infinitely many consequences, it is bound to be explanatory for some of these. As already mentioned, the distinction between principle and constructive theories cannot be sharp. Nevertheless, we would not want to say that 'The letter is under the cup on the table' is an explanation of anything, at least not in the first instance—neither, for Einstein, is special relativity.

Following Norton (1991) we might say that Einstein's verificationism insists that *theories should not distinguish between states when there is no observable difference between them.* I think Norton's version is very nearly the right way to express Einstein's position, but I want to modify it slightly to incorporate thought experiments and say that Einstein's principle of verification is:

> *Theories should not distinguish between states when there is no* intuitive *difference between them.*

The crucial difference between 'observable' and 'intuitive', as I use these terms here, is this. An intuitive distinction is one that can be made either in ordinary experience or *in a thought experiment.* (Thus, an observable difference is just a special case of an intuitive one.) This is to be contrasted with a (purely) theoretical distinction, one made in a (constructive) theory, which cannot be 'perceived' even in a thought experiment.[4]

We can see the spirit of Einstein's verificationism at work in the thought experiment described in the opening paragraph of the special relativity paper. Einstein's brand of verificationism is different from the more traditional empiricist sort which insists on sticking to observable elements only when doing any sort of theorizing. Einstein, on the contrary, is happy with all sorts of unobservable things. It also differs from any view that says there is no fact-of-the-matter to distinguish theories which are observationally equivalent. Two theories might make exactly the same empirical predictions, but they can still be distinguished on the basis of what is pictured to be happening in a thought experiment.

The magnetic induction example from the beginning of the special relativity paper perfectly illustrates this. There is no intuitive difference (i.e., no observable difference in the thought experiment) between the conductor moving while the magnet is at rest and the magnet moving while the conductor is at rest. (Yet there is a theoretical difference—the presence or absence of an electric field.) So only their relative motion should be taken into account.

> It is known that Maxwell's electrodynamics—as usually understood at the present time—when applied to moving bodies, leads to asymmetries which do not appear to be inherent in the phenomena. Take, for example, the reciprocal electrodynamic action of a magnet and a conductor. The observable phenomena here depend only on the relative motion of the conductor and the magnet, whereas the customary view draws a sharp distinction between the two cases in which either the one or the other of these two bodies is in motion. For if the magnet is in motion and the conductor is at rest, there arises in the neighbourhood of the magnet an electric field with a certain definite energy, producing a current at the places where parts of the conductor are situated. But if the magnet is stationary and the conductor in motion, no electric field arises in the neighbourhood of the magnet. In the conductor, however, we find an electromotive force, to which in itself there is no corresponding energy, but which gives rise—assuming equality of relative motion in the two cases discussed—to electric currents of the same path and intensity as those produced by the electric forces in the former case.
>
> (Einstein 1905a, 37)

Einstein then goes on to say 'Examples of this sort . . . suggest that the phenomena of electrodynamics . . . possess no properties corresponding to the idea of absolute rest' (1905a, 37). The principle of relativity is then 'raised to the status of a postulate' (1905a, 38).

The crucial thing to note here is that Einstein does not rail against either fields or currents, neither of which are observable. In fact, in the magnetic induction example not only is the observable needle motion the same in both cases, but the unobservable current is the same in both cases, as well. The phenomena are identified as being the same phenomenon in both cases; in other words, there is no distinction to be made in the intuitive aspects of the thought experiment, so our electrodynamic theory must adjust itself to this fact.[5]

Einstein and Leibniz

It has often been claimed that Leibniz gave verificationist arguments against absolute space. It may prove instructive to contrast Einstein with Leibniz, since their respective brands of verificationism may be similar. In a very beautiful and highly influential thought experiment, Leibniz imagines God creating the material universe in different places in Newton's absolute space. He then objects:

(a)

I   Conductor moves, magnet is at rest. Motion of conductor through magnetic field causes current, which causes needle to move in the ammeter.

(b)

II   Magnet moves, conductor at rest. Motion of magnet (changing magnetic field) causes electric field to exist, which causes current, which in turn causes the needle to move.

Figure 25

'tis impossible there should be a reason why God, preserving the same situations of bodies among themselves, should have placed them in space after one certain particular manner, and not otherwise; why everything was not placed the contrary way, for instance by changing East into West. But if space is nothing else, but the possibility of placing them; then those two states, the one such as it now is, the other supposed to be the quite contrary way, would not at all differ from one another. The difference therefore is only to be found in our chimerical supposition of the reality of space in itself. But in truth the one would be the same thing as the other, they being absolutely indiscernible; and consequently there is no room to enquire after a reason of the preference of the one to the other.

(Alexander 1956, 26)

For Leibniz the indiscernibility which is at the heart of the issue is not mere *observable* indiscernibility, but some sort of complete indiscernibility. That is, all the theoretical apparatus in the world can be brought to bear on the question, and still there would be no way to distinguish two universes which are East-West reversed. In my terminology, the different ways God might have created the universe are intuitively alike; there is no observable or intuitive difference between them, even in a thought experiment. Leibniz is no ordinary empiricist—his verificationism covers the results of thought experiments, too.

Einstein's brand of verificationism is much closer to Leibniz than to positivism. To see the difference let us, following Michael Friedman,[6] contrast the Leibniz-Einstein brand of verificationism with that of a true positivist, Moritz Schlick, who was one of the first philosophers to comment on relativity. In his account of relativity which links it to positivism, Schlick writes:

points which coincided at one world-point x_1, x_2, x_3, x_4 in the one universe would again coincide in the other worldpoint x'_1, x'_2, x'_3, x'_4. Their coincidence—and this is all we can observe—takes place in the second world precisely as in the first . . . . The desire to include, in our expression for physical laws, only what we physically observe leads to the postulate that the equations of physics do not alter their form in the above arbitrary transformation . . . . In this way Space and Time are deprived of the 'last vestige of physical objectivity,' to use Einstein's words.

(1917, 53)

Unlike Schlick's, the Leibniz-Einstein brand of verificationism is one with which a realist can happily live. Of course, a realist about space-time will be unhappy, but theoretical entities are not all ruled out in principle, as they would be on Schlick's brand of verificationism. In each case, atoms, fields, or space-time will have to be argued for or against as the situation warrants. As it turns out, atoms and fields are permissible as far as Einstein is concerned, but space-time, perhaps, is not. A true positivist would eliminate them all. Someone who insists only on intuitive differences is more discerning.

Real experiments

Ilse Rosenthal-Schneider once asked Einstein a famous 'What if . . .' question.

> Suddenly Einstein interrupted the reading and handed me a cable that he took from the window-sill with the words, 'This may interest you.' It was Eddington's cable with the results of the famous eclipse expedition. Full of enthusiasm, I exclaimed, 'How wonderful! This is almost the value you calculated!' Quite unperturbed, he remarked, 'I knew that the theory is correct. Did you doubt it?' I answered, 'No, of course not. But what would you have said if there had been no confirmation like this?' He replied, 'I would have had to pity our dear God. The theory is correct all the same.'
>
> (1980, 74)

Of course, it is easy to adopt such a confident stance when victorious; but what if things really had gone a different way? There is an example when the experimental outlook did not seem so good for Einstein; this is in the case of the Kaufmann experiments which were widely interpreted as refuting special relativity.

Kaufmann was a skilled experimenter working largely in the tradition known as the 'electromagnetic view of nature'. This was a school of opinion which flourished in the late nineteenth and early twentieth centuries. The central idea was that electromagnetism, not mechanics, is the real foundation of physics. Perhaps the most profound claim was that mass itself is electromagnetic in origin, being the result of a charged body interacting with the electromagnetic field. In such a context it would then be quite natural to ask whether mass varied with velocity, and if so, to what extent? Distinctions between longitudinal and transverse mass, which would be nonsense in classical mechanics, are perfectly meaningful here. Kaufmann performed a series of experiments on the relation between mass and velocity (see Miller 1981 for details), and the results were unfavourable to both Einstein and Lorentz. In the very early days, the two theories of Einstein and Lorentz were often identified. And Kaufmann rejected them together on the basis of his data, since their predictions for mass variation with velocity were at odds with his experimental findings. Lorentz's reaction is interesting. In a letter to Poincaré he writes, 'Unfortunately my hypothesis of the flattening of electrons is in contradiction with Kaufmann's results, and I must abandon it' (quoted in Miller 1981, 334). (Wouldn't Popper be pleased!) Poincaré was almost as pessimistic as Lorentz. For him the principle of relativity was an experimental fact, but he was prepared to dump it.

Einstein, however, largely ignored Kaufmann's experimental results. Why? Holton paints a picture of the victory of a great theory over experience.

> With the characteristic certainty of a man for whom the fundamental hypothesis is *not* contingent either on experiment or on heuristic (conventionalistic) choice, Einstein waited for others to show over the next years, that Kaufmann's experiments had not been decisive.
>
> (1964, 190)

In another place Holton says that Kaufmann's experiments mark

> the crucial difference between Einstein and those who make the correspondence with experimental fact the chief deciding factor for or against a theory: even though the 'experimental facts' at that time very clearly seemed to favor the theory of his opponents rather than his own, he finds the ad hoc character of their theories more significant and objectionable than an apparent disagreement between his theory and their 'facts'.
>
> (1973, 235)

These two passages contain much insight. Nevertheless, I think they both miss the target. There is nothing in Einstein's work to suggest he really thought the principle of relativity or the constancy of light postulate were 'not contingent on experiment'. He quite clearly states the contrary. And second, when Holton says that Einstein was put off by the *ad hoc* character of other theories, he is taking into account features of theorizing which have to do with constructive theories. When Einstein talked of 'inner perfection' (e.g., absence of *ad hoc* features) being a requirement, this is a requirement of a constructive theory. The epistemological status of relativity, as Einstein repeatedly stressed, has quite a different character.

In contrast with Holton's theory-overrules-experience interpretation of Einstein's reaction to Kaufmann, I suggest instead that it was a battle of one kind of experience *vs* another kind of experience. It was not a case of clinging to a bold conjecture in the face of conflicting observations, but rather a case of clinging to one class of experiences (derived in part from thought experiments and embodied in the principle of relativity and the constancy of light postulate) in the face of apparently conflicting experimental observations. As I said earlier, there is no sharp distinction between principle and constructive theories, so I cannot claim here that it was a clear case of observation *vs* observation instead of theory *vs* observation. Nevertheless, Einstein's opposition to Kaufmann was much more like a case of one class of observations in conflict with another.

Of course, Einstein was vindicated, but in the context of 1906 was he just a stubborn fool who got lucky? For Poincaré and Lorentz the theory being tested by Kaufmann was (in Einstein's terms) a constructive theory—and they quite rightly abandoned it in the face of contrary empirical evidence. But if we understand special relativity as a principle theory then hanging on in the face of Kaufmann may have been exactly the rational response.

Einstein and Bohr

So far I have argued that: (1) there was no pilgrimage for Einstein from Machian empiricism to some sort of realism. (2) Einstein maintained his (peculiar form of) verificationism throughout his life, and it is sometimes connected to thought experiments and always tied up with what he called 'principle theories'. (3) Einstein simultaneously maintained an H-D, or conjectural methodology which was linked to what he called 'constructive theories'. (4) It is quite possible that Einstein was never at any time a genuine realist, but this is largely irrelevant to the methodological issues at hand. (5) Understanding Einstein's reaction to any 'refuting' experiment should always be viewed in the light of the principle/constructive distinction. (6) The illusion of a change in Einstein's philosophical outlook is largely due to a shift in his scientific interest from relativity, a principle theory, to quantum mechanics, a constructive theory.

Now it is time to say something about his attitude toward quantum mechanics. As is well known, Einstein intensely disliked it. Many were very surprised by his rejection. I have already mentioned Heisenberg, who was shocked that Einstein did not accept a line of reasoning (Only allow observables!) which Heisenberg took to be central to the success of relativity. Max Born, who despaired of ever getting Einstein on side, remarked that 'He believed in the power of reason to guess the laws according to which God has built the world' (1956, 205). Einstein's deepest debate was with Bohr; it went on for years, but neither side budged. (See Bohr 1949 for a history of their debate.)

On the 'pilgrimage' view of Einstein, given by Holton, Fine, and others, Einstein's rejection of quantum mechanics is not in the least surprising: Einstein had become a realist, so he rejected quantum mechanics' self-imposed restriction to observable elements and he instead posited a hidden reality which lay behind the phenomena.

I want to suggest a different way of looking at things. On my view, Einstein would have been just as unhappy with quantum mechanics had he seen it in 1905 as he was in 1925 or 1955. Quantum mechanics is a constructive theory, and Einstein never had qualms about positing a hidden realm to explain the phenomena (e.g., in 1905, molecules to explain Brownian motion).

When people express surprise that Einstein, whom they think to be a positivist, would not accept the (apparently) same line of reasoning in quantum mechanics as he did in relativity, they miss something important. But, on the other hand, by merely calling the mature Einstein a 'realist', as Holton does, and using this to account for Einstein's opposition to quantum mechanics, Holton and others overlook a vagueness in the concept of 'realism'. The idea contains at least two distinct features.

If there was a change in Einstein's philosophical views from Machian empiricism to realism, then the difference was largely epistemological. To be a realist in this regard is to think we can have rational beliefs about a non-observable realm. It is in this sense that, for example, van Fraassen (1980) is an anti-realist, since he is a sceptic about anything non-observable. This is *not* what is at issue in Einstein's quarrel with Bohr.

The other sense of realism is much more concerned with ontological or metaphysical issues; it is contained in the idea that the truth of a theory (or a single sentence) is *independent* of theorizers.[7] It is in this latter sense of realism that, say, Kant is an anti-realist. For him the truth of 'A causes B' or 'X is left of Y' fundamentally depends on rational agents. This, according to Kant, is the way we necessarily conceptualize our experience. However, there is in reality no causation, nor are there spatial relations among things-in-themselves.

The fight between Einstein and Bohr was over this ontological aspect of realism. On the so-called Copenhagen interpretation of quantum mechanics, a measurement does not discover the magnitude of some system; rather it *creates* the result. Until a position measurement is made, an electron, for example, has no position at all. The Heisenberg uncertainty principle says in effect that if a position measurement is made, then a position is created but there is no momentum at all. It is not that there is a momentum and we cannot know what it is; rather, there simply is no momentum. It was the violation of this aspect of realism which so troubled Einstein. On Bohr's view (as Einstein saw it) the micro-world does not exist *independently* of our theorizing about it.

It is one thing to say Einstein developed from a Machian empiricism to accepting theoretical entities and the legitimacy of HD methodology; but the battle against Bohr and quantum mechanics was more like a battle against Kant, a battle against the view that nature is dependent on us.[8] 'Physics', remarks Einstein, 'is an attempt to conceptually grasp reality as it is thought independently of its being observed' (1949, 81). Such an outlook is perfectly compatible with the brand of verificationism I am attributing to him. Though I reject the developmental view anyway, even if it were correct it still would not explain Einstein's attitude toward quantum mechanics.

Einstein is also famous for rejecting the alleged indeterminism of quantum mechanics. 'God does not play dice.' But the question of determinism is also independent of any verificationist *vs* H-D methodology debate.

Perhaps I have overstated things. A sharp distinction between epistemology and ontology is at the heart of the realist outlook. If Einstein had been a thoroughgoing Machian, the distinction would not have been appropriate to him. So it might be said that only by becoming a realist could Einstein make the appropriate distinction which in turn enabled him to criticize Bohr and other champions of the quantum theory.

This strikes me as a plausible view, and to some extent it is still in the spirit of the developmental account's version of events. Perhaps, then, the appropriately cautious thing to say in concluding this section is simply this. If we are looking for a nice, neat, straightforward explanation of Einstein's rejection of quantum mechanics, we will *not* get it from his (alleged) development from empiricism to realism. I would prefer to look in an entirely different direction, though I am not sure just where. Of course, this agnosticism applies only to the decade 1925–35; after that, his reason for opposition to quantum mechanics is manifestly obvious. By the mid-1930s we have the EPR thought experiment which yields the simple *principle* theory: The magnitudes of physical systems exist independently of any measurement. The quantum theory is constructive, so no talk of its wonderful empirical success can stand up to a principle theory in conflict with it, since the latter, after all, possesses 'logical perfection and security of foundation'.

Concluding remarks

I began by expressing dissatisfaction with several aspects of the developmental view. Even though Holton's 'pilgrimage' story has a happy ending—Einstein breaks free from appearances, marries reality, then rides off into the sunset—there are just too many lacunae in the story to make it believable. In its place I have given an account which takes his verificationism seriously and finds a role for thought experiments in it; I see it as enduring throughout his scientific career and as doing very valuable work. But I also see it as compatible with a general realist outlook: even if Einstein was not himself a thoroughgoing realist, we should be.

There is perhaps a methodological lesson to be learned from Einstein. Normally realists hold some sort of H-D or broadly conjectural view of *all* theorizing. Theories are conjectures and are believed true because of their consequences (i.e., explanatory power, novel predictions, etc.). Of course, in Einstein's terminology, these are constructive theories. For principle theories, like relativity, the story is quite a different one. Principle theories are fallible, but they nevertheless have quite a different feel about them than do the bold conjectures of his constructive theories.

I have used Einstein's own distinction between principle and constructive theories, not because I think historical characters should be allowed to tell their own stories—far from it. Rather, I have used it because it seems to capture a real distinction in genuine theorizing, a style that Einstein himself practised so well. The onus is now, perhaps, upon us to spell out in greater detail the difference between principle and constructive theories and the workings of Einstein's particular brand of verificationism. It served Einstein well, once; it may serve the rest of us well again.

8 Quantum mechanics
A Platonic interpretation

For more than seven decades quantum mechanics (QM) has been a philosophical nightmare. All sorts of good sense have been dashed on the rocks of this brilliantly bizarre bit of physics. So, I am not too sanguine about the prospects of the proposal I'm about to present. Nevertheless, in the spirit of 'nothing ventured, nothing gained' I shall outline the philosophical problem (or at least one of the main problems) then try to offer a solution. The proposal is a Platonic one, linked to what has been done in earlier chapters. If it works it will support the earlier Platonic account of laws and thought experiments. If it doesn't, then I can only hope it is at least an interesting failure.

The road to Copenhagen

The makers of QM understood their formalism in a rather straightforward, classical way. If the initial ideas of Schrödinger or Born had worked we wouldn't have the philosophical problems that we do have today.[1] Erwin Schrödinger, for example, thought of the $|\psi\rangle$ in his equation as representing a physical entity, say, an electron. It was conceived to be a wave, more or less spread out in space. However, this view faced enormous problems. For one thing, $|\psi\rangle$ can spread over a great expanse, but we always measure electrons as point-like entities; for another, $|\psi\rangle$ is a complex function and in some cases many-dimensional, so it cannot be a wave in ordinary three-dimensional space. These and other problems made the initial Schrödinger interpretation hopeless. So, even though the Schrödinger equation is a central ingredient in QM, $|\psi\rangle$ must be understood in quite a different way than he intended.

Max Born proposed that $|\psi\rangle^2$ should be understood as the probability density for the location of an electron. On his view the electron is a particle; the state vector $|\psi\rangle$ just tells us the probability amplitude of it being located at various places. The waves of so-called wave-particle duality are probability waves—they are a reflection of our ignorance, not of the physical world itself which is made of localized particles. Born's view was also philosophically attractive in that it made QM look like classical statistical mechanics—there is a world which exists independently of us; it's just that we cannot have a complete description of it. Alas, Born's view worked no better than Schrödinger's. It ran afoul of interference effects. Consider the split screen experiment. We have an electron gun, a barrier with two small openings, and a detecting device such as a photographic plate. The pattern on the back screen with both slits open is *not* the sum of the patterns with only one slit open at a time. There is interference. We can even slow down the firing rate so that there is only one electron in the device at a time, thus undermining the explanation that electrons are bumping into one another. Consequently, the only reasonable conclusion is that the electron is interfering with itself, just as a wave might do. It would seem, *contra* Born, that |ψ⟩ is not *just* a reflection of our knowledge and our ignorance, but has some sort of physical reality to it.

We can think of Schrödinger's view as an ontological interpretation of QM, since |ψ⟩ is about the world, and Born's as an epistemological interpretation, since |ψ⟩ is about our knowledge of the world. Since neither of these philosophically straightforward interpretations works, the attractiveness of the Copenhagen approach becomes somewhat inevitable.

The Copenhagen interpretation is mainly the product of Niels Bohr, though there are numerous variations. Bohr thought the wave and particle aspects of any physical system are equally real; |ψ⟩ has both ontological and epistemological ingredients. As Heisenberg put it, 'This probability function represents a mixture of two things, partly a fact and partly our knowledge of a fact' (1958, 45). A state of superposition is not a mere state of ignorance—*reality itself is indeterminate*. An electron, for example, does not have a position or a momentum until a position or a momentum measurement is made. In classical physics, observations *discover* reality, but in QM, according to Bohr, they somehow or other *create* the world (or at least the micro-world).

Before going any further, let's look closely at the formalism of QM, the very thing of which we are trying to make sense.

The QM formalism

The standard theory consists of a number of principles. The following is a simple version of the postulates and some of the main theorems of orthodox QM (in one dimension).

1. A physical system is represented by a vector $|\psi\rangle$ in a Hilbert space. All possible information is contained in $|\psi\rangle$.
2. Observables, A, B, C, . . . (e.g., position, momentum, spin in the z direction, etc.) are represented by Hermitian operators, **A, B, C,** . . . each with a complete set of eigen-vectors, $|a_1\rangle$, $|a_2\rangle$, $|a_3\rangle$, . . ., $|b_1\rangle$, $|b_2\rangle$, $|b_3\rangle$, . . ., etc.
3. Measurements result in eigenvalues only; i.e., a_1, a_2, a_3, . . . (which correspond to the eigenvectors $|a_1\rangle$, $|a_2\rangle$, $|a_3\rangle$, . . .). If a system is in state $|\psi\rangle$ and A is measured, then

$$\text{Prob}_{|\psi\rangle}(a_i) = |\langle a_i|\psi\rangle|^2 = |a_i|^2.$$

4. The following three propositions are equivalent:

    (1) **A** and **B** are compatible observables.
    (2) **A** and **B** possess a common set of eigenvectors.
    (3) **A** and **B** commute.
         Moreover, when **A** and **B** do not commute (i.e., **AB–BA** = $c \neq 0$), then $\Delta\mathbf{A}\,\Delta\mathbf{B} \geq c/2$ (the uncertainty principle).

5. The state of the system evolves with time according to the Schrödinger equation

    $$H|\psi\rangle = i\hbar d|\psi\rangle / dt$$

6. After an interaction, two systems, 1 and 2, with their states in Hilbert spaces H_1 and H_2, are represented by the state $|\psi_{12}\rangle$ which is in the tensor product space, $H_1 \otimes H_2$.
7. If a measurement of observable A results in a_i then right after the measurement the system is in the eigenstate $|a_i\rangle$.

A few words of explanation. The first two postulates link physical systems to mathematical structures. A physical system, say an electron, is represented by a vector in an infinite dimensional vector space, called a Hilbert space. The properties that the system can have, e.g., position, momentum, angular momentum, energy, spin, etc. are called observables (though often there is nothing observable about them) and are associated with operators. Examples in one dimension are: the position operator: $X = x$, and the momentum operator: $P = -i\hbar d/dx$. Operators are functions defined on the Hilbert space. For example, if $|\psi\rangle = e^{ipx/\hbar}$ then we have $P|\psi\rangle = i\hbar d(e^{ipx/\hbar})/dx = (-i\hbar)(ip/\hbar)e^{ipx/\hbar} = pe^{ipx/\hbar}$.

Whenever an operator, O, has the effect of just multiplying a vector by a real number r, the vector is called an eigenvector of that operator. That

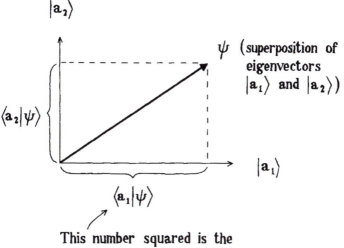

$|a_2\rangle$

ψ (superposition of eigenvectors $|a_1\rangle$ and $|a_2\rangle$)

$\langle a_2|\psi\rangle$

$|a_1\rangle$

$\langle a_1|\psi\rangle$

This number squared is the probability of obtaining the value a_1 when ψ is measured.

Figure 26

is, if $O|\psi\rangle = r|\psi\rangle$ then $|\psi\rangle$ is an *eigenvector* and r is called an *eigenvalue*. According to the third principle, only eigenvalues can be the results of any measurement. In the simple example just given, p is an eigenvalue and will be the magnitude revealed on measurement. (At least in the ideal case—it is acknowledged by all that real life is messy.)

The state of the system might correspond to one of the eigenvalues, but it need not in general; often it will be in a state of *superposition*, a linear combination of two or more eigenvectors (which are the basis vectors of the space). The extent to which $|\psi\rangle$ overlaps some eigenvector gives the probability amplitude of getting the corresponding eigenvalue as a result of measurement.

Generally, the state of the system will change as time passes. This is represented by having $|\psi\rangle$ move around in the Hilbert space in a manner governed by the Schrödinger equation. This change of state is smooth and deterministic.

When two quantum systems interact they remain 'entangled' in some sense, even when far removed from one another. The tensor product of two such states is not a simple product of these two states considered separately. The combined state has some remarkable properties of its own. The whole, as we shall see, is greater than the sum of its parts.

The final principle is perhaps the most controversial; it is known as *the projection postulate* and it says that measurements bring about discontinuous, non-deterministic changes of state, transitions from superpositions to eigenstates. It is controversial because it looks as if observation does not merely discover reality—but somehow creates it. More of this below; but first, some general remarks about what the philosophy of physics tries to do.

Interpretations

What is even meant by 'an interpretation of the QM formalism' is somewhat vague. Logicians have a precise notion of 'interpretation' or 'model of a formal system', but that won't do here. To start with, the formalism is already partially interpreted; it is hooked to observational input and output in a clear and unambiguous way. This partial interpretation is called the minimal statistical interpretation. What it can do is handle everything observable. It is often favoured by those who advocate an instrumentalist outlook for scientific theories in general, sometimes expressed as 'Shut up and calculate'. But our interest is with how the world really works, not just with making successful observable predictions. Only those lacking a soul are content with the minimal statistical interpretation.[2] What's needed is something over and above this instrumentally adequate, but otherwise incomplete account.

The spectrum of possible interpretations is exceedingly broad. On the one side it may be as trivial as so-called hidden variables. (That is, trivial in the philosophical or conceptual sense since a hidden variables interpretation tries to make the quantum world out to be as much like the classical world as possible. Such interpretations are not trivial in the technical sense—indeed, they are probably impossible.) On the other hand, the range of possible interpretations may be limited only by our imaginations. The empirical consequences of rival views are largely unknown. It was long thought that the realism involved in the EPR thought experiment made no observable difference—hence it was often branded 'idle metaphysics'. But Bell, to the surprise of everyone, derived an empirically testable consequence. One could even imagine consequences of the right interpretation of QM being as significant as special relativity, for it is quite plausible to see Einstein's theory as an interpretation of Maxwell's electrodynamics.

I'll broadly consider two types of interpretation of QM here. *Realist* interpretations of QM hold that the quantum world exists independently of us; we do not create it in any way; quantum systems have all their properties all the time; measurements discover those properties, they do not create them. On the other hand, *anti-realist* views, such as the Copenhagen interpretation, hold that the quantum world is not independent of us; in some important sense observers make reality; measurements create their results.

The minimal statistical interpretation is a kind of anti-realism, too, but it's important to note the difference. Scientific realism typically involves at least two ingredients—epistemic and ontological. The first says that we can have rational beliefs about a realm of unobservable entities. Opposition to this aspect of realism is a form of scepticism. Duhem and van Fraassen both share this form of anti-realism with the minimal statistical interpretation of QM. On the other hand, neither Kant nor Bohr are sceptics about unobservable entities, but they alike reject the second aspect of scientific realism which holds that the world we are trying to learn about exists *independently* of us.

From Bohr to Wigner

The problem largely boils down to making sense of superpositions or entanglements. According to Bohr, superpositions are not mere states of our ignorance. When an electron, for example, is in a state of superposition between two position eigenstates (as it is in the split screen experiment), it does not really have one position or the other. This state corresponds to having both slits open. The closure of one slit corresponds to a precise position measurement which would in effect create a position for the electron.

Heisenberg's famous uncertainty principle (e.g., for position and momentum we have $\Delta p \, \Delta q \geq \hbar$ is to be understood ontologically: if we measure for position, say, then we create a position and leave the momentum uncreated. This is not to be confused with an ignorance interpretation of the uncertainty principle which says we merely don't know what the momentum is. For Bohr, there simply is no momentum. (To reflect this ontological construal, many prefer 'indeterminacy principle' to 'uncertainty principle', since the latter may misleadingly suggest an epistemic interpretation.)

So how does the quantum system change from a state of superposition to an eigenstate? Bohr's answer to this, his so-called relational interpretation, is somewhat involved. Bohr held to a sharp distinction between the micro- and the macro-worlds. QM is true of the former and classical physics true of the latter. I stress *true*, not a mere approximation; trees, tables, and trucks do not go into states of superposition for Bohr.

The wave function of a micro-entity collapses, according to Bohr, when it is in the right sort of relation with a macro-object. Thus, an electron goes into an eigenstate of position in the split screen device when one slit is open; a photon goes into a spin eigenstate in, say, the x direction, when it passes through an appropriately oriented polaroid filter. In each case there is the right sort of relation between micro-and macro-object to bring about the collapse of the wave function. Notice that a conscious human observer need not become aware of the results, but the existence of the macroscopic world is required.

Though sympathetic with Bohr's outlook in general, Eugene Wigner (1962) and others could not follow Bohr in his micro/ macro distinction. There is only one physical world, according to Wigner; so if QM is right anywhere, it is right everywhere. Thus the distinction between micro and macro, while epistemologically significant, is ontologically unimportant. Indeed, it is certainly standard practice among physicists to treat every-day objects as though they are correctly described by QM. In almost every text one first learns the de Broglie relation, $p = h/\lambda$, and then as an exercise one is asked to compute the wavelength of a truck roaring down the highway. Of course, the moral of the exercise is that the wavelength is very small, even undetectable, but the tacit implication of such cases is unmistakable—QM applies to the big as well as the small.

A unified account is naturally very satisfying, but it comes at a cost. It means that the macroscopic devices that are used to measure quantum systems are themselves QM objects, subject to states of superposition. It was this that Schrödinger made so much of in his cat paradox thought experiment that we discussed in the first chapter—the macroscopic cat is in a superposition of living cat and dead cat.

Suppose, however, that we are prepared to bite the bullet and allow—absurd though it seems—that macroscopic objects such as measuring devices can go into superpositions. How then can a measurement even take place? What collapses the wave function?

Consider an energetic atom, A, that has some probability of decay during some time period. In general, its state will be a superposition of the two eigenstates, the energetic state $|e\rangle$ and the decayed state $|d\rangle$.

$$|\psi_A\rangle = |e\rangle + |d\rangle.$$

Now let's bring in a geiger counter, G, to see whether it decays or remains energetic. If the atom decays, it will flash; but if there is no decay, then no flash. (This is idealized to the point that G makes no mistakes, which is, of course, quite unrealistic. But we shall continue to ignore this.) Let these two eigenstates of G be $|f\rangle$ and $|n\rangle$, for the flashed and non-flashed eigenstates, respectively. We now have a combined system on our hands; it is represented by the tensor product of the two systems:

$$|\psi_{A+G}\rangle = |e\rangle \otimes |n\rangle + |d\rangle \otimes |f\rangle.$$

Hence, we now have the geiger counter which is not in any definite state, but is rather in something like a state of superposition of its two eigenstates, flash and no-flash. How do we collapse its wave function? We don't have a measurement of the atom while G is in a superposition, so how is a measurement even possible? We could bring in another measuring device, say a TV camera—but that would only push the problem back a step. How about someone just looking? This wouldn't seem to help either, since the human eye, optic nerve, and brain are all physical objects and are thus correctly described by QM. Consequently, even our physical brains go into states of superposition. There seems no hope of stopping this regress. The combined system consisting of atom, geiger counter, TV set-up, human eye, and brain appears forever destined to be in limbo. Yet, we do make measurements; we do get eigenvalues. But how?

Wigner's solution is a radical mind/body dualism. He argues as follows. What we need to collapse wave functions is some sort of entity or process that is non-physical, not subject to quantum mechanical superpositions. Do we already know any candidates for this? Indeed we do—the *mind*. Human consciousness, says Wigner, is the thing which can put physical systems into eigenstates. The mind is not a physical entity; it is not subject to superpositions. (A bad hangover doesn't even come close.) Consciousness can do the one thing no physical entity can do—turn the indeterminate into the real.

Neither Bohr's nor Wigner's account of measurement is very satisfactory for a number of reasons, not the least of which is that each must go *outside* QM for the essentials of the account: Bohr to an independent macroscopic world, Wigner to consciousness. In classical mechanics, on the other hand, the theory of measurement is itself classical and part of the theory. Ideally, the right theory of measurement in QM would be from QM—anything else smacks of the *ad hoc*.

EPR

The anti-realism of the Copenhagen interpretation, whether Bohr's or Wigner's version, was met head-on by the beautiful EPR thought experiment. It was discussed briefly above, but we need a bit more detail now.

The argument proceeds by first characterizing some key notions.

Completeness: A theory is complete if and only if every element of reality has a counterpart in the theory. Thus, if an electron, for example, has both a position and a momentum, but the theory only assigns a value to one and not the other, then that theory is incomplete.

Criterion of Reality: If, without disturbing the system, we can predict with probability one the value of a physical magnitude, then there is an element of reality corresponding to the magnitude. The qualification—without disturbing the system—is central. The Copenhagen interpretation holds that measurements do disturb the system (they collapse the wave function), so ascribing an independent reality to any magnitude cannot be based on a (direct) measurement.

Locality: Two events that are space-like separated (i.e., outside each other's light cones) have no causal influence on one another. They are independent events. This follows from special relativity which holds that nothing, including causal connections, travels faster than light.

The more perspicuous Bohm version of the EPR argument starts with a system such as an energetic particle that decays into a pair of electrons; these travel in opposite directions along the z axis. Each electron, call them L and R (for left and right), is associated with its own Hilbert space. The polarization or spin eigenstates will be along any pair of

Figure 27

orthogonal axes, say, x and y, or x' and y'. In any given direction a measurement (which, recall, only yields eigenvalues) will result in either +1 for the spin up state or –1 for the spin down state. We can represent these as $|+\rangle_L$ and $|-\rangle_L$, respectively, for the L electron, and $|+\rangle_R$ and $|-\rangle_R$ for R. The spin of the system is zero to start with and this must be conserved in the process. Thus, if L has spin magnitude +1 in the x direction then R must have –1 in the same direction to keep the total equal to zero. A composite system such as this, in the so-called singlet state, is represented by

$$|\psi_{LR}\rangle = 1/\sqrt{2}\,(|+\rangle_L \otimes |-\rangle_R + |-\rangle_L \otimes |+\rangle_R)$$

If we measure the spin of the L electron we then know the state of R since the measurement of $|\psi_{LR}\rangle$ immediately puts the whole system into one or other of the two eigenstates. Suppose our measurement resulted in L being polarized in the x direction (i.e., has spin up in the x direction). This means the state of the whole system is $|+\rangle_L \otimes |-\rangle_R$, from which it follows that the remote electron is in state $|-\rangle_R$, i.e., it has spin down in the x direction. (Choosing the x direction is wholly arbitrary; any other direction could have been tested for.) While it might be conceded that the measurement on L may have 'disturbed' it, or created rather than discovered the measurement result, the same cannot be said of R. We are able to predict with complete certainty the outcome of the measurement on the R electron, and since (by the locality principle) we could not have influenced it in any way with a measurement of L, it follows (by the criterion of reality) that the R magnitude exists independently of measurement. Since this is not reflected in $|\psi\rangle$, it follows (by the criterion of completeness) that QM does not completely describe the whole of reality. EPR then concludes that the theory must be supplemented with hidden variables in order to give a full description.[3]

Schrödinger's kittens

We can make the EPR result even more compelling by considering distant correlations of macro-systems. Here's a modification of Schrödinger's cat paradox which I'll call the kitten paradox. Assume Schrödinger's cat had two kittens. After a suitable passage of time we separate the kittens from the poison and from each other. All this will be done without looking at them, i.e., without measuring the state of their health. Thus each, while being transported, will be in a state of superposition of living and dead just as in Schrödinger's original thought experiment. The only difference is that we now have two and they are moved outside of each other's light cones before being observed.

Let the distant observers now examine each kitten. I confidently predict both will be in the same state of health. (Let's say they are both alive. One of the joys of thought experiments is that one can sometimes stipulate a happy ending.) This perfect correlation suggests they were already both alive before being separated.

Observation merely revealed what was already the case; it had no influence on the kittens' health. Describing the kittens as being in a state of superposition—not really alive or dead—cannot be a completely correct description of reality.

The kitten example is just EPR on a macroscopic scale. The result—that measurements *discover* the state of the kittens' health—seems a common-sense truth, too trivial to tell. Alas, if only it were.

The Bell results

Compelling though EPR is, it can't be right. This is the upshot of several related findings known collectively as the Bell results. Bell's original argument was rather complicated, but versions are now so simple that those with only elementary algebra can easily comprehend the argument. I'll begin with a simple derivation of a Bell-type inequality (due to Eberhard 1977), then briefly describe its experimental refutation.

Let us begin by considering an EPR-type set-up. Unlike EPR, however, we will consider measurements of spin in different directions, say along a and a' for the L photon and b and b' for R. There are four possible measurements that could be made:

$$(a, b), (a', b), (a, b'), (a', b')$$

(where (a, b) means the L photon is measured for spin along the a axis and R along the b axis). A spin up result of a measurement has value $+1$, and spin down -1. Now define a correlation function, $c(x, y)$ as follows:

If $a = 1$ and $b = 1$, then $c(a, b) = 1 \times 1 = 1$;
if $a = 1$ and $b = -1$, then $c(a, b) = 1 \times -1 = -1$;
and so on for a', b', etc. (where a=1 means that the result of measuring the L photon in the a direction is $+1$, etc.)

We imagine running the experiment many times. After N tests, with a_i being the ith result, we have

$$c(a, b) = 1/N \Sigma_i a_i b_i.$$

We will make two key assumptions.

> *Realism*: Each photon has all of its properties all of the time; in particular, each has a spin up or spin down magnitude in every direction whether there is a spin measurement made in that direction or not.

This assumption is embedded in the mathematics as follows. Let a_i (or a'_i, b_i, b'_i, respectively) be the result of the ith measurement if made in the a (or a', b, b', respectively) direction. The value is either +1 or −1 and this value exists *whether a measurement is made or not*. In particular, if photon L is measured in the a direction then it cannot be measured in the a'direction. Nevertheless, even though we can't know what it is, we still assume that it has one value or the other. This is the core of realism—measurements do not create, they discover what is independently there.

> *Locality*: The results of measurement on one side of the apparatus do not depend on what is happening at the other side. The outcome of a spin measurement on photon L is independent of the direction in which R is measured (i.e., the orientation of the apparatus); it is independent of the outcome of that measurement; and it is independent of whether R is measured at all.

Formally, the locality assumption is captured by having the value of a_i be independent of the values of b_i and b'_i. So if a measurement of L in the a direction would result in +1 if R were measured in the b direction, it would still be +1 if R were measured in the b' direction instead. Recall that Bohr holds that a micro-entity has its properties only in relation to a macro-measuring device; different settings may create different micro-properties. Locality does not completely deny this, but it does deny that the settings of a *remote* macro-device have any influence.

Now define the following formula which I'll call F for convenience:

$$F = a_i b_i + a_i b'_i + a'_i b_i - a'_i b'_i$$

Rearranging terms we have

$$F = a_i(b_i + b'_i) + a'_i(b_i - b'_i)$$

Since the a terms equal +1 or −1, and since one of the terms in parentheses equals 0 while the other equals either +2 or −2, we have

$$F = +2 \text{ or } -2$$

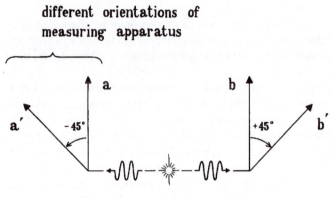

different orientations of measuring apparatus

source

Figure 28

Thus, taking the absolute value, we have

$$|a_i b_i + a_i b'_i + a'_i b_i - a'_i b'_i| = 2$$

This holds for the ith measurement result. The generalization for N measurements is therefore

$$|1/N \sum_i a_i b_i + a_i b'_i + a'_i b_i - a'_i b'_i| < 2$$

In terms of the correlation function we have

$$|c(a, b) + c(a, b') + c(a', b) - c(a', b')| \leqslant 2$$

This is one form of Bell's inequality. It means that when spin measurements are done for arbitrary directions a and a' on the L photons and b and b' on the R photons, we can expect that degree of correlation. After many tests the correlations between the L and R photons, taken a pair at a time, must satisfy this inequality—*if* the assumptions of realism and locality both hold.

It is important to stress that the inequality is derived by a simple combinatorial argument based on two common-sense assumptions: realism and locality. QM, however, makes a different prediction. An experimental test of QM and Local Realism (as it is often called) is thus possible.

To get specific QM predictions we need to specify directions for the spin measurements to be made. Let a = b, otherwise the orientations of a and b can be arbitrary; furthermore, let a' be –45 degrees and b' be +45 degrees from the common a/b direction. According to QM the correlation functions have the following values:

$$c(a, b) = -\cos 0 = -1$$
$$c(a, b') = -\cos 45 = -1/\sqrt{2}$$
$$c(a', b) = -\cos 45 = -1/\sqrt{2}$$
$$c(a', b') = -\cos 90 = 0$$

What this means is that if L is measured in the a direction and has, say, spin up, then R measured in the b (=a) direction will not have spin up. They are perfectly negatively correlated. In the fourth case immediately above when L and R are measured at right angles to each other the results of measurement are completely uncorrelated. The other two cases yield results in between.

We now substitute these values derived from QM into the Bell inequality:

$$|-1 -1/\sqrt{2} -1/\sqrt{2} - 0| = 1 + 2/\sqrt{2} > 2$$

Thus, at these angles, QM and Local Realism diverge in their predictions, making an empirical test possible.

Just how remarkable this situation is cannot be stressed too much. For years EPR was attacked by the empiricist-minded for being 'idle metaphysics' since it was thought that it made no detectable difference. Even defenders of EPR were willing to concede that the realism/anti-realism debate has no empirical import. It was thought to be *merely* a thought experiment. Now it turns out that all were wrong. The thought experiment has a real experiment analogue. Abner Shimony calls it 'experimental metaphysics', and the phrase is exactly right. From the original EPR argument to Bell's derivation of his inequality, to the experimental tests, to the reaction to those tests, there is an inextricable mix of physics and metaphysics. The whole situation is highly reminiscent of the seventeenth century, when philosophy was at its very best.

If the possibility of performing an empirical test on realism was surprising, the outcome was even more surprising—common sense has taken a beating. There have been several tests of the inequality. In almost every one, QM has made the right predictions and Local Realism the wrong ones. Of all these tests, the ones carried out by Aspect *et al.* (1981, 1982a, 1982b) have been the most sophisticated.

The crucial feature of the Aspect experiment is the presence of a very fast optical switch which directs L photons to either a or a' and R

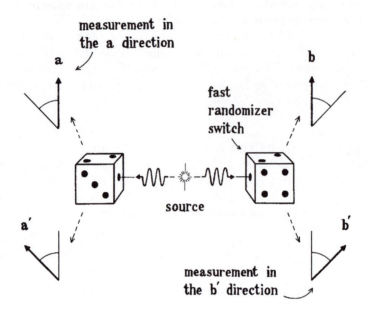

measurement in
the a direction

a

b

fast
randomizer
switch

source

a′

b′

measurement in
the b′ direction

Figure 29

photons to either b or b′ measurements. It picks a direction randomly, while the photon is in flight. The reason this is considered important is that in earlier experiments the setting of the distant measuring device was fixed long before the measurement, thus allowing the possibility of a sub-luminal signal between the distant wings of the apparatus and hence the possibility that they could 'communicate' with one another. Of course, that may seem bizarre, but the QM world is so weird that it is always nice to have one more possibility ruled out, however far-fetched it may seem to common sense.

Concepts of locality

The very notions of locality and action at a distance are somewhat com-plicated. A few words seem called for. Historically[4], *action at a distance* has meant the causal influence of one material body upon a second that is separated from the first by a void. Two bodies gravitationally attracting one another with only empty space between them is an example. Even if there were a medium between the distant bodies, we still would have action at a distance, if the medium played no role in the action of one body upon the other. The contrary notion is *local action*; it occurs when one body acts on a second which is in contact with it or when there is a medium between the bodies that conveys the action.

Descartes held that all action of one body on another is imparted by contact or by pressure transmitted through an etherial medium (thus, all action is ultimately by contact). In spite of its elegance and numerous achievements, the so-called mechanical philosophy and its near relations had serious problems, such as explaining the cohesion of bodies. Why do things hold together? Any new explanation that proposed some sort of action at a distance, such as an attractive force, was taken to be a retrograde step, a return to the despised occult qualities of the medievals.

Newton seemed to many to be guilty of this when he introduced universal gravitation. He was roundly criticized by latter-day Cartesians and by Leibniz, who, for example, asserted that any attraction across *empty* space that caused a body to move in a curved line could only be a miracle (which he took to be absurd).

Newton seemed to agree. In one of his famous letters to Richard Bentley, he asserts

> that one body may act upon another at a distance through a *vacuum*, without the mediation of anything else, by and through which their action and force may be conveyed from one to another, is to me so great an absurdity, that I believe no man who has in philosophical matters a competent faculty of thinking, can ever fall into it.
>
> (Cohen 1978, 303)

Many of the action at a distance problems with the mechanical philosophy were laid to rest with the rise of field theory. According to Michael Faraday, for example, a field of force surrounds an electrically charged particle. This field pervades all space and is, strictly speaking, part of the actual body. Consequently, any body is really in direct contact with every other body, since each, in a sense, occupies the whole universe.

Faraday gave a very convincing argument for the reality of these fields. The interaction that is conveyed by the field has a finite velocity. Add to this the principle of the conservation of energy. Now if one body is jiggled, the second will respond after some time interval. Where is the jiggling energy located during the intermediate times? It cannot be located in either body, so it must be located in the field. Thus, the field is a real thing.

This argument is the basis for calling special relativity a field theory. By making all velocities of interaction finite, special relativity, in a sense, implies the existence of fields. Of course, it doesn't follow that if we deny an upper bound on the velocity of interaction we thereby deny the existence of a field or medium that transmits the action, but special relativity is so deeply linked to field theory that asserting the existence of a superluminal connection has become almost synonymous with asserting action at a distance. And even if there is no medium that carries the action from one body to another, as long as that action is conveyed at a velocity equal to or less than c, then the action is called 'local'. The issue of action at a distance has evolved considerably since its heyday in the seventeenth century, but I think this is the link that connects them.

Within QM in general 'action at a distance' means action transmitted faster than the speed of light, the upper bound allowed by special relativity; and 'local action' means action transmitted at speeds equal to or less than the speed of light. Still there is lots of room for details and variations inside QM itself. One of the most thorough treatments of the issue can be found in Redhead (1987), and I can do no better than follow his account.

Redhead starts out by positing a general locality principle.

L: Elements of reality pertaining to one system cannot be affected by measurements performed at a distance on another system.

He then discusses several different interpretations of QM and formulates locality principles suitable for each.

LOC_1: An unsharp value for an observable cannot be changed into a sharp value by measurements performed at a distance.

LOC_2: A previously undefined value for an observable cannot be defined by measurements performed at a distance.

LOC_3: A sharp value for an observable cannot be changed into another sharp value by altering the setting of a remote piece of apparatus.

LOC_4: A macroscopic object cannot have its classical state changed by altering the setting of a remote piece of apparatus.

LOC_5: The statistics (relative frequencies) of measurement results of a quantum mechanical observable cannot be altered by performing measurements at a distance.

Redhead goes on to characterize two more principles of locality. (Both have technical expressions; I'll merely quote his gloss, though even that contains some terms undefined here. Remember, I'm just trying to convey the spirit of complexity which surrounds these issues.)

Ontological Locality: locally maximal observables on either of two spatially separated systems are not 'split' by ontological contextuality relative to the specification of different maximal observables for the joint system.

Environmental Locality: the value possessed by a local observable cannot be changed by altering the arrangement of a remote piece of apparatus which forms part of the measurement context for the combined system.

Much of Redhead's book is devoted to carefully determining the precise conditions under which each of the various locality principles is violated. I shall be much less subtle—no doubt at my peril—and try to give general arguments against rival views lumped together.

Options

The derivation of Bell's inequality made an empirical test of Local Realism *vs* QM possible—Local Realism lost. Of course, no test is really crucial; perhaps there is a way to wriggle out and save the two common-sense assumptions of realism and locality. That, however, seems unpromising. It appears that we must give up at least one of these two assumptions.

Abandoning premises is hardly the end of the affair, however; we still have the perfect correlations of the EPR case to explain. So what are the options?

1. A brute fact

This view hangs on to locality—since we don't want to abandon special relativity—but it gives up realism. As for the EPR correlations, this view says: 'It's just a brute fact of nature; that's simply the way things are'. In short, there is no explanation for the EPR correlations; they are just a brute fact. (Van Fraassen 1980, 1982 seems to hold this view.)

In many (perhaps all) scientific theories, there are elements which are taken as just brute facts. For instance, in Newton's physics, inertia is an unexplained explainer; it accounts for other phenomena, but is itself unaccounted for. Are EPR correlations like that? I think not. It is one thing to make the mass and charge of quarks, say, a basic, primitive, unexplained feature of reality, according to our theory; but it is quite another to say it is impossible in principle to ever explain these properties. Yet this is what the (empiricist) brute fact view does to EPR correlations. By clinging to locality and denying realism it makes any sort of future explanation impossible in principle. Though logically possible, the brute fact view is so unattractive that we should adopt it only as a last resort.

In a paper which is interesting both historically and philosophically, Don Howard (1985) argues that Einstein distinguishes between *separability* ('spatially separated systems possess their own separate real states' (1985, 173)) and *locality*. He claims that this distinction is at the heart of Einstein's objections to Bohr, and that the point was lost in the EPR paper. (Einstein was unhappy with the way Podolsky had written up the argument in EPR.)

Howard's philosophical claim is that the Bell inequality can be derived from the two assumptions of locality and separability, and (in light of the experimental results) that we should abandon separability, not locality. Spatially separated quantum systems which have previously interacted are not really distinct—there is really only one system. This, according to Howard, is what quantum holism amounts to. (Howard's distinction is similar to distinctions made by Jon Jarrett (1984)—completeness *vs* locality—and by Abner Shimony (1986)—outcome independence *vs* parameter independence.)

It is very hard to get a grip on such an idea. Even if it is only 'one system', does it not have a left side and a right side? And how do these sides interact? Given the existing correlations, do not the two sides of the 'one system' have to interact at superluminal speeds? Such questions may be ruled out of court, but it leaves the idea of non-separability (coupled with locality) something of a mystery. As Howard himself puts it: 'to say we accept non-separable quantum theory is not to say that we yet understand all that such acceptance entails' (1985, 197). Until it is clarified, I'm inclined to lump this account with the brute fact view. I hope I'm not as naive as Lord Kelvin who said he couldn't understand a theory until he had a mechanical model of it, but I think we should insist on some sort of mechanism for non-separability. And when we have it, how can it avoid non-locality without doing something quite absurd like abandoning space-time?

2. Non-local, contextual hidden variables[5]

Hidden variable theories come in two main types. Non-contextual theories simply attribute sharp values for all observables at all times. They have been ruled out in principle by a deep mathematical result known as Gleason's theorem. However, the actual hidden variables theories which have been developed have often been contextual. According to a contextual theory the results of measurement depend on two things: the state of the system (which includes the hidden variables) and the environment (in particular, the state of the measuring apparatus).

Local hidden variable theories (whether contextual or not) are ruled out by the Bell results. Non-contextual hidden variable theories (whether local or not) have been ruled out by Gleason's theorem. Non-local, contextual hidden variable theories remain a possibility. David Bohm's (1952) account is probably the paradigm example of such a non-local, contextual hidden variable theory.

On this view, realism is true but locality is false. The magnitudes that a physical system has depend on the whole physical situation within which the system finds itself. In particular the magnitudes depend (in part) on the setting of the remote wing of the whole experimental set-up. This view is similar to Bohr's in that it is holistic, being both non-local and relational. But it differs from Bohr's in stipulating that a micro-system has all its properties all the time.

In order for the view to work there must be some sort of (physical) causal connection between the photons and the whole measuring apparatus, especially some connection with the distant parts of the apparatus. (Bohm's 1952 hidden variable theory, for example, posits an all-pervasive 'quantum potential'.) This connection will have to be superluminal, which at once raises a problem. What about special relativity? There are two choices.

2(a) Special relativity false. This is such a drastic move that it should be avoided at almost all costs. There are those who are prepared to make it (Bell 1987, Popper 1982), but I don't think we should join them; special relativity plays too great a role in the rest of physics.

2(b) Peaceful co-existence. On this view it is noted that the superluminal signals that are responsible for the correlations are completely uncontrollable. This fact is alleged to save special relativity. At one side of the EPR-type set-up we get a random sequence of spin up and spin down results. We know that the other side is getting a down result when we get up (and up when we get down), but there is no way to control this sequence of ups and downs. We cannot, for instance, use this random sequence to synchronize distant clocks and thereby undermine the relativity of simultaneity.[6]

Champions of this view thus declare that there is a peaceful co-existence between QM and special relativity. In contrast with 'action at a distance', Abner Shimony, who advocates the view, calls it 'passion at a distance'. It deserves extensive consideration. After criticizing it I'll turn to a third option.

3. *Laws as causes of correlations*

This option takes laws of nature—understood to be entities existing in their own right—as non-physical causes of quantum events. It will be explained below.

Passion at a distance

I have two objections to the 'peaceful co-existence' or 'passion at a distance' view. The first is philosophical; it concerns motivation.

The point is to save realism, but this view saves realism in QM by undermining it in space-time. No longer can relativity be considered a theory of an objective space-time; it must now be viewed as a theory about our operations, about what signals we can or cannot detect. On the passion at a distance interpretation, special relativity has a decidedly operationalist flavour; it is a theory of what we can or cannot control rather than a theory about an independent space-time. So the price of getting realism into QM is losing it in the space-time realm—hardly an advance.

My second objection is somewhat technical and rather speculative. What I shall try to show is that any non-local hidden variable account (with special relativity taken to be true) implies a different statistics than that implied by the singlet state.

First, an assumption. A crucial premiss—frankly, this is the fuzzy bit—is that there must be some sort of mechanism, some sort of signal transmitted at the time of measurement which links the two particles, the two wings of the singlet state, and which accounts for the correlated eigenvalues. It might, for example, be a disturbance or a change in potential in the 'field' of David Bohm's (1952) non-local hidden variable theory. A second assumption is that there are two ways of breaking the holism of the singlet state. One way is to measure a particle directly; the other is to measure its correlated, distant partner. This assumption seems reasonable since a particle can be detected in an eigenstate either directly or by having its mate measured. Either way, when the holism of the singlet state is broken (or when the state vector 'collapses', as champions of the Copenhagen interpretation would say), it lets the world know by sending out a superluminal signal to all and sundry. Think of this as a kind of action-reaction similar to what would happen in a gravitational field with two massive particles—jiggle one and a force is sent out which wiggles the other, which in turn sends back a force which will wiggle the first, etc., etc.

Now consider a standard EPR set-up, with one significant difference. Let L and R be the frames of the left and right particles (as well as their names), with L stationary and R in relative motion with velocity v. It is the relative motion of the two wings of the experiment that is unusual. Normally they are stationary. At the time of interaction with the left measuring device, L sends a superluminal signal to R with velocity u. (In the limit this can be infinitely fast.) This sets the spin components and also breaks the holism of the singlet state. The distance from L to R is d, and $c = 1$ throughout. (See Figure 30, which is a centre of mass diagram that is perspicicuous but does not correspond strictly to the derivation.)

R's co-ordinates are

$$t_R = \frac{t_L - vx_L}{\sqrt{(1-v^2)}} \text{ and } x_R = \frac{x_L - vt_L}{\sqrt{(1-v^2)}}$$

When R breaks free from the holistic singlet state, a signal is sent out with velocity

$$-u = \frac{dx_R}{dt_R}$$

Since $dt_R = \dfrac{dt_L - vdx_L}{\sqrt{(1-v^2)}}$ and $dx_R = \dfrac{dx_L - vdt_L}{\sqrt{(1-v^2)}}$

we can substitute and solve to find the velocity in L. This gives

$$-\frac{dx_L}{dt_L} = \frac{u-v}{1-uv}$$

The total time $= \Delta t = $ (distance/velocity out) + (distance/velocity back). Thus,

$$\Delta t = d/u + d/[(u-v)/(1-uv)]$$

Since $u >> c = 1$ we have $t < 0$. In the limit, as $u \to \infty$ and $v \to 1$, $d/u \to 0$ and $(u-v)/(1-uv) \to -1$; hence the right hand side approaches $d-1$; thus, we get a total time of $-d$. This, of course, means that the signal was received *before* it was sent. (This argument is modelled on one showing tachyons go backward in time.)

This implies that the holism of the singlet state was broken before a measurement was made. Speaking in the language of the projection postulate, we could say that the wave function thus collapsed earlier; hence a signal was sent out to its partner putting it into an eigenstate earlier, and so on back into the more distant past, right back to the origin. Speaking in more realistic language, we could say that it was wrong to describe the state $|\psi_{LR}\rangle$ as being the entangled tensor product of the Hilbert spaces of L and R. In any case, *the singlet state cannot exist in any spatially extended system, given the assumptions of special relativity and the existence of a non-local mechanism.* The L and R systems would not be correlated as they are in the entangled singlet state; they would have quite different measurement outcomes, namely, the same as predicted by local realism (i.e., satisfying the Bell inequality). However, the experimental results that actually do occur are compatible only with those represented by the singlet state, that is, they would violate Bell's inequality. No EPR-type experiment with one wing of the aparatus in motion has ever been performed, but we can be confident of the outcome. So, non-local, contextual hidden variables are experimentally refuted. Grand conclusion: Peaceful co-existence is experimentally refuted. Contextual hidden variables are no more possible than Local Realism.

Let me anticipate a possible objection. These superluminal signals carry no energy or momentum, which makes them quite unlike ordinary

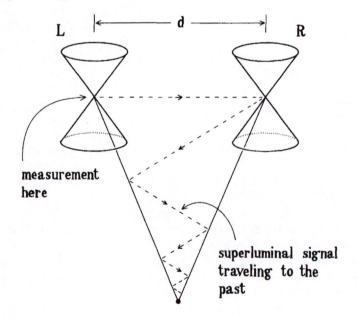

Figure 30

signals. It also makes them quite unlike tachyons; however, they do have a kinematical similarity. Perhaps we could have a 're-interpretation principle' similar to that which is popular among tachyon theorists. This is a principle which is used to try to get around causal paradoxes: A superluminal signal which is *absorbed* after going backwards in time can be *re-interpreted* as one which is being *emitted* and going forward in time. Such a reinterpretation may help save tachyons, but it won't help here. Whether coming or going, the mere presence of a superluminal signal means that the system has broken free of the holism of the singlet state.[7]

Correlations and the laws of nature

So far we have seen how the EPR argument called for Local Realism, but that the Bell results have made that impossible. However, that can't be the end of the matter, since we still have the simple EPR correlations to account for. All we know from the Bell results is that Local Realism can't be the right explanation. We have also seen that attempts to develop non-Local Realism have not been satisfactory either. A different kind of account of QM is needed.

There are a number of conditions or desiderata that should be satisfied. (1) An acceptable account of QM should allow for an explanation of the correlations that do exist. We should not pass off the perfect correlations of the EPR case, for example, as mere coincidences or as just brute facts of nature. (2) It should not violate special relativity. There should be no *physically* non-local mechanisms involved. To these two requirements I wish to add another. (3) An interpretation of QM should also take into account the fact that experimenters have *immediate knowledge* of measurement outcomes at the remote wing of the apparatus, as a result of measurements on the near side. Thus, an account which can do justice to the existence of our knowledge as well as to the physical aspects of the problem is to be preferred over an account which does justice only to the first two conditions.

My suggestion is very simple. *Distant correlations are caused by the laws of nature.* I realize this sounds almost silly. One wants to say the correlations *are* the laws of nature. But this, if you recall preceding chapters, is not true. A law of nature is an independently existing abstract entity—a thing in its own right which is responsible for physical regularities. The spin correlations that do occur in the physical world are brought about by a non-physical law. It is this very same entity, the abstract law, which plays a role in our knowledge of what is going on at the distant wing of an EPR-type experiment.

On this interpretation realism is true; every QM system has all its properties all the time. Locality, on the other hand, is false in the sense that there is a statistical dependence of values on the remote wing of the apparatus; but there are certainly no physical causal connections, that is, no so-called efficient cause. In some respects this view is similar to contextual hidden variables, but there is one crucial difference—there are no superluminal signals. It is not a physical connection, but abstract laws which are the cement of the universe (to use Hume's expression). I am perfectly happy to call this a cause, but it is akin to a *formal cause* as Plato or Aristotle would have understood it.

Earlier I referred to Howard's 'non-separability' (and the similar notions of Jarrett and Shimony) as a brute fact view of quantum correlations. Perhaps an alternative way of viewing this idea is by linking it to the abstract laws—laws are the (non-physical) mechanism which hold 'non-separable' quantum systems together. Understood this way, Howard's view is not so far from my own. In spite of the fact that they seem to have non-local effects, rather than calling laws of nature 'non-local' I prefer to call them 'global', since even though they are transcendent, they are, in another sense, everywhere. If an atheist may be allowed the simile, laws of nature are like God—omnipresent. These are the things that account for the antics of micro-entities.

Conclusion

How does all this relate to thought experiments? My account of Platonic thought experiments assumes the existence of laws of nature construed as entities in their own right. They are not physical things, but they are perfectly real and completely independent of us. The present chapter has utilized laws of nature, so understood, to provide an interpretation of QM. In so far as this has seemed a plausible view, it then supports the realist view of laws, which in turn supports (albeit indirectly) my Platonic account of thought experiments. But the relation is actually more intimate than this picture of indirect support suggests. In the EPR thought experiment, which has been the heart of all this, the source of the correlation has been the same as the source of the thought experimenter's knowledge of the remote measurement results—namely, a law of nature. There is a remarkable harmony here: the structure of knowing and the structure of reality mirror one another. Or as Spinoza put it: The order and connection of ideas is the same as the order and connection of things' (*Ethics,* II, Prop. 7).

Afterword

I began with a variety of illustrations of thought experiments in the first chapter, and in the second proposed a system of classification. The taxonomy included a special type, which I called Platonic. Before arguing for the *a priori* character of these very remarkable thought experiments the stage was set with a chapter on Platonism in mathematics. Those who give some credence to mathematical realism and who take seriously our intuitions of mathematical reality will find that it is not so big a step to a Platonic account of (some) thought experiments. At any rate, this is what I have tried to argue: In a small number of cases, thought experiments give us (fallible) *a priori* beliefs of how the physical world works. With the mind's eye, we can see the laws of nature.

Of course, laws of nature, as objects to be seen, cannot be what every empiricist thinks them to be—supervenient on events. Rather, laws of nature must be things in their own right. It turns out that the Platonic account of thought experiments dovetails very nicely with the realist view of laws proposed by Armstrong, Dretske, and Tooley, and I am very happy to adopt their account.

The whole theory—the *a priori* epistemology of thought experiments and the realistic metaphysics of laws—was then used to look at Einstein's scientific practices and at the interpretation of quantum mechanics. If the theory is believable, then this is due in part to the direct arguments for it given in chapter five, in part to the success of the last two chapters on Einstein and quantum mechanics, i.e., it is believable because of its good consequences, and in part with its ties to visual reasoning in mathematics.

That is a brief account of what I have tried to do; now a few words about what I haven't attempted. There are a number of issues concerning thought experiments that certainly merit future consideration.

Kant received almost no attention in this work. This is mainly because there was no existing Kantian position on this topic until very recently.

There is a large body of writings on 'constructability', a notion which is undoubtedly relevant to thought experiments, but there was no Kantian treatment of the classic examples. Marco Buzzoni (2008) has recently filled the gap and will undoubtedly become a serious rival to the various existing empiricist and Platonic accounts.

Natural kinds were briefly discussed near the outset when it was conjectured that we can draw powerful general results from a single thought experiment because the object of the thought experiment is taken to be an instance of a natural kind. I dare say the interconnections between thought experiments and natural kinds are rich and varied, and certainly worth pursuing. A start can be made with the works I mentioned earlier (i.e., Wilkes 1988 and Harper 1989). In some respects the work of James McAllister is closely related (McAllister, 1996, 2004, 2005).

Computers are used increasingly for discovery purposes. Of course, since the beginning, computers have been used for computations, but their uses in modelling phenomena, a use which often sheds enormous light on whatever situation might be at hand, makes them a profound tool in the (non-algorithmic) process of discovery. Much of this computer modelling looks remarkably like thought experimenting. What, precisely, is the relation?

The actual number of questions that may arise from a study of thought experiments is probably limitless. It's a fascinating and fertile field. In the first edition I said that I hoped others would cultivate it. It turns out that many did. It has not been possible for me to take up the many (sometimes brilliant) ideas, new points of departure, or outright criticisms of me. The field has grown enormously and a new edition such as this can only fill in a few things here and there. What is really needed is a new book, starting from scratch. This I hope to do, but it won't be for a while. In the meantime, I would like to mention a few of the major works that have appeared since the first edition of this book and to also mention a few of the themes that I think should be addressed.

Of course, there were major historical works on thought experiments, by Mach and by Kuhn, for instance, but relatively little consideration was given until the late 1980s and early 1990s. As well as the first edition of this book there was the anthology edited by Horowitz and Massey (1991) and books by Sorensen (1992), Haggqvist (1996), Gendler (2000), Rescher (2005), and Buzzoni (2007). For German readers, three important books are: Genz (1999), Kühne (2005), and Cohnitz (2006). Kühne's book is a very thorough history. Highly influential recent articles have been those by Norton, Nersessian, Miščević, McAllister, Gendler, and Gooding, to name but a few. (See Bibliography for details.) Their work has been the point of departure for much subsequent work on the topic.

There is also a large and growing body of literature by those who do not have a particular interest in the sciences. Philosophers who think of themselves as epistemologists or as metaphysicians sometimes deal with modal reasoning or the nature of intuition. There is an obvious overlap with the work of philosophers of science concerned with thought experiments. A good sample of this sort of work can be found in the anthologies edited by Gendler and Hawthorne (2002) and by DePaul and Ramsey (1998), and the recent book by Williamson (2007). I dare say there is much to be learned by each side from the other, but a gap remains and it will remain until the work of non-philosophers of science can be shown to be relevant to the best example of knowledge that we have—science.

The central problem of thought experiments is simply put: How is it that we seem able to learn about the world just by thinking? I think everyone would agree with that. Where they disagree is in the range of answers, some even contending that the 'seeming' to learn about the world is actually a delusion. Beside the big question, there are several particular ones. I won't try to present them as *the* outstanding questions, but rather as merely some of the questions that interest me.

1. Why do some fields such as physics and philosophy have so many thought experiments (which are hugely important to those fields), while others such as chemistry have so few or none at all? Any answer to this question is likely to tell us something important about the field itself.

2. Literary fictions (novels, plays, movies) have a narrative structure similar to a thought experiment and they often teach us lessons of some sort or other. Is this similarity only superficial or does it run deep? (See Davies (2007), Elgin (2007), Oatley (1999), and Swirski (2007).)

3. Does culture and background matter? I often try out the Galileo thought experiment involving the rate of falling objects on students with varying degrees of training in science. They are almost all persuaded by the example. Would this be true of people in very different cultures? There has been lots of anthropological and psychological work on moral reasoning in different cultures using thought experiments, but none (as far as I know) on physics examples. Culture matters to some extent in the moral case; does it also matter in the physics cases?

4. The legitimacy of thought experiments might vary from field to field. Most moral philosophers use thought experiments in their work, but a minority consider them highly misleading and quite illegitimate. So-called 'What if . . . history' (e.g., what if the Nazis had made the atom bomb first?), is decried by many historians but others find it highly useful. Who is right?

5. To what extent are thought experiments and visual reasoning in mathematics similar to and as legitimate as thought experiments in physics? Scepticism is the norm, but there are some striking cases that might lead us to reconsider visual reasoning as more fuitful and reliable than usually thought. I presented a brief sample in chapter four, but much more can be said. (See Brown (2008) for a more detailed sympathetic presentation and Giaquinto (2007) for a mixed account.)

These are just a few of the many questions about thought experiments that could and should be pursued. Based on what I have seen at recent conferences, I would say that the field is flourishing. The number of people involved in growing, new proposals abound, and the quality of work is very high. Many areas of contemporary philosophy seem moribund and scholastic, but the field of thought experiments is certainly not one of them.

Notes

1 Illustrations from the laboratory of the mind

1 From the title page of Stevin's *Wisconstige Gedachenissen*, more commonly known by its Latin translation *Hypomnemata Mathematica*, Leyden 1605.
2 I owe the initial observation to David Papineau; that Kant and the constructive interpretation of geometry should be stressed I owe to Kathleen Okruhlik.
3 For example, Strawson (1959), in a philosophical thought experiment, considers a world consisting entirely of sounds.
4 Ian Hacking (1983) has stressed the manipulation or intervention aspect of (real) experimentation. Many of his observations carry over to thought experiments. David Gooding (in conversation) stressed the common 'procedural' nature of real and thought experiments—a perceptive characterization.
5 For the sake of the argument it is assumed the cat does not observe itself; otherwise we can simply change the example to some non-conscious macroscopic entity in place of the cat.
6 Thomson's paper contains other interesting thought experiments as well. As I said, Thomson accepts that the fetus has a right to life *only* for the sake of the argument. Michael Tooley (1984) uses several interesting thought experiments to argue that the fetus does not have a right to life.
7 The Turing test, devised by Alan Turing, says: If we cannot distinguish between humans and a computer when interacting with them (over a teletype so that we cannot actually see what we are interacting with) then the computer truly can be said to think.
8 Fundamental articles by Locke, Hume, Williams, Parfit, and many others on the topic of personal identity are conveniently reprinted in Perry (1975). This literature is loaded with ingenious and amusing philosophical thought experiments.
9 Igal Kvart (in conversation) suggested making the nomological distinction the basis of legitimacy and stressed that brains-in-a-vat are nomologically possible while splitting people are not. The idea (but not the example) is the same as Wilkes's, though she uses the expression 'theoretical possibility'. Since we're on the issue, I should also mention that some ethicists have criticized the thought experiments, too, saying that the danger is chiefly in abstracting from considerations that really are morally relevant, namely, social and political context.

10 I have Polly Winsor and Michael Ruse to thank for the biological examples and especially for disabusing me of my belief that biology is impoverished when it comes to thought experiments.

2 The structure of thought experiments

1 This taxonomy is based on the one given in my 1986 publication, but there are some small changes introduced here.

2 But not exactly momentum, since Descartes eschewed mass. For him 'quantity of motion' would be more like 'size of matter times speed'. Leibniz didn't have a clear notion of mass either, though unlike Descartes he was not wedded to a purely kinematic physics.

3 Leibniz's thought experiment started the so-called *vis viva* controversy. For an account of the history as well as a discussion of historians' views of that controversy see Laudan (1968) and Papineau (1977).

4 As I mentioned earlier, John Norton (1991) has a fundamental division between deductive and inductive thought experiments. But I don't think this is where the action is at all.

5 In what immediately follows I owe much to Wilkes (1988), which links natural kind reasoning to thought experiments, and to Harper (1989), which is a suggestive discussion of the relation between inductive inference and natural kinds in general.

6 As I mentioned in the previous note, Wilkes also thinks that natural kinds play an important role in thought experiments. According to her (1988, 12f.) reasoning about natural kinds is what allows thought experiments in physics to be successful. On the other hand, 'persons' are not natural kinds, which is in part why personal identity thought experiments are illegitimate. However, 'person' seems to me a straightforward natural kind. I much prefer Wilkes's other objections: we don't have the relevant background information; the possible world in which people split like an amoeba is too remote from our world for us to make any sensible or plausible claims about it.

4 Mathematical thinking

1 I am indebted to Alasdair Urquhart for pointing out the following example and for helpful discussions about it. Of course, this doesn't mean he would agree with all my claims about it.

2 Feferman (1983) and Wang (1974, 1987) both contain very enlightening discussions of Gödel's Platonism.

3 There is no intention here of adopting a serious observational/ theoretical distinction.

4 Kitcher (1983) defeats *a priori* accounts of mathematics by unfairly identifying *a priori* knowledge with *certain* knowledge.

5 Whether Field's programme is successful is debatable; for a critical discussion see Urquhart (1989).

6 This objection may not be fair since Maddy distinguishes perceptual from inferential knowledge. She writes, 'I base my case that the belief that there are three eggs is non-inferential on empirical studies that suggest we don't count, don't infer, for such small numbers' (private communication). I am not as

confident as she is that we have a sharp distinction, but I will not pursue the issue here. I am grateful to her for this and many other clarifications of her view.

7 Maddy (1980) makes this and several other excellent points in her critique of Chihara (1973).

8 I am grateful to David Savan for telling me of this example.

9 Harman (1973) is one prominent example. Goldman (1967), selections from Harman, and other useful discussions can be found in Pappas and Swain (1978).

5 Seeing the laws of nature

1 Otherwise I could perform the thought experiment now and derive 'The moon is made of green cheese'.

2 See McAllister (1989) for an insightful and provocative account of aesthetics in science.

3 The term 'knowledge' may be too strong as it implies *truth;* 'rational belief might be better since, onmy view, what is *a priori* could be false.

4 Following this one Hume gave two other definitions which he seemed to think were equivalent. They aren't, but they are in the same regularity spirit: 'if the first object had not been, the second never had existed.' And 'an object followed by another, and whose appearance always conveys the thought to that other'.

5 In Tooley's version (1977), which is quite Platonic, the universals have a transcendental character. By contrast, Armstrong's version (1983) is somewhat naturalistic; it does not allow the existence of uninstantiated universals. I have a strong preference for Tooley's platonism.

6 I am much indebted to Demetra Sfendoni-Mentzou for directing me to Peirce and for providing me with a great deal of information about his views of laws of nature.

7 Real in the sense of being in the final science. Thus, laws are just as real as trees or electrons. On the other hand, nothing is real for Peirce in the sense of so-called metaphysical realism.

8 This claim needs some qualification. Probabilistic evidence is obviously not transitive. The principle stands, however, for the examples at hand.

9 Most Kuhnian theses are disturbing to empiricists, but this one might prove relatively attractive since it tries to solve the problem of how we can learn something about nature without making any new observations.

10 I am much indebted here to John Collier's forthcoming paper, 'Two Faces of Maxwell's Demon Reveal the Nature of Irreversibility'.

11 Many physicists are loath to admit that QED is inconsistent; they will claim that the re-normalization problem has been solved. I won't argue the matter here; but this much should be admitted by all: *Prior* to the work of Feynman, Schwinger, and Tomanoga in the late 1940s QED was inconsistent. Moreover, classical electrodynamics always was and remains inconsistent. In both cases we have reasoning from inconsistent premises.

12 Stalnaker has posited an inconsistent world to be the place where contradictions hold. It's a useful fiction, not unlike a 'point at infinity', which makes the sematical machinery run smoothly. However, it's no help here, since, first, all consistent worlds are closer, and second, every proposition

holds there so we can't tell what the 'legitimate' predictions of QED really are.

13 Unlike some logicians with an interest in 'para-consistent logic', I do not for a moment believe the world is inconsistent. An inconsistent theory is certainly false, but it may nevertheless be better than some of its consistent rivals.

14 To cite another extreme, Paul Feyerabend (1975) makes Galileo out to be a fellow epistemological anarchist. A spectrum of views can be found in the Butts and Pitt collection (1978).

6 The development of inertial motion

1 The distinction between actual and potential infinity is central to all so-called constructive accounts of mathematics. It remains a contentious issue to this day. Brouwer, for instance, will only accept the potential infinite. The majority of mathematicians, however, cheerfully accept the existence of actual infinities. For a brief discussion, see Brown (2008).

2 Some go so far as to say, 'It can scarcely be doubted that impetus is analogous to the later inertia, regardless of ontological differences' (Clagett 1961, 524). There is some justice to this so long as one ignores the relativity of motion. But once relative motion is introduced the similarities begin to fade.

7 Einstein's brand of verificationism

1 I must sadly confess that I'm increasingly less sure of this.

2 Philipp Frank reports a similar encounter in which Einstein, to Frank's complete surprise, rejected any sort of positivism (1947, 214f.).

3 I mean H-D to be understood very broadly: a theory is tested by its observable consequences. Thus, Popper, Lakatos, Laudan, and Bayesians are H-D methodologists.

4 Cheryl Misak and I exchanged works in progress and were surprised to find similar formulations of verificationism—I for Einstein; she for C.S.Peirce. I strongly recommend her forthcoming book on the subject, *Truth and the Aims of Inquiry* (Oxford).

5 Here again I am indebted to Norton (forthcoming) both for the style of the diagram and to a large extent for the analysis of this thought experiment.

6 I am much indebted here to Michael Friedman (1983) which perceptively discusses both Leibniz and Schlick in connection with Einstein.

7 Discussions of the different senses of 'realism' can be found in Newton-Smith (1982), Horwich (1982), and Putnam (1983).

8 This brief description is far from doing justice to either Bohr or Kant. Kant, for example, was an 'empirical realist' and stressed the independence (in one sense) of the observer from the observed. But ultimately, the properties of the 'phenomena' depend on our conceptualization. My analogy between Kantian and Bohrian senses of 'dependence' is only a loose one.

8 Quantum mechanics: a platonic interpretation

1 For a thorough account of early interpretations of QM (as well as present-day ones) see Jammer (1974).

2 Of course this is quite unfair. Consider the remark a bit of gentle polemics. Many who hold the minimal view (reasonably) feel forced into it by the repeated failures of realistic approaches to QM.

3 The usual reading of the EPR paper sees it as calling for hidden variables. However, some recent commentators have denied this and suggested that a weaker thesis is being asserted. See Fine (1986, 26–39).

4 For an excellent historical survey of these involved issues see Hesse (1961).

5 See Shimony (1984) for a discussion of various aspects.

6 There are proofs (e.g., Eberhard 1978 and Jordan 1982) that distant correlations can never upset the random nature of quantum events; thus, it would be impossible to send a 'message' faster than c.

7 I am much indebted to Harvey Brown for an extensive discussion of this and related issues. Among other things he pointed out some possible problems with standard tachyon thinking which may have a bearing on my argument since algebraically they are very similar. For a review of the relevant tachyon literature see Recami (1986). I also greatly profited from comments by Kent Peacock who made me clarify and correct some important points. I fear this argument will prove to be controversial, and neither Harvey Brown nor Kent Peacock should be understood as endorsing it.

Bibliography

Abbott, E.A. (1952) *Flatland*, New York: Dover.

Alexander, H.G. (ed.) (1956) *The Leibniz-Clarke Correspondence*, Manchester: Manchester University Press.

An Anonymous Treatment of Peripatetic Dynamics, in Clagett (1961).

Aristotle, *Physics*, in Clagett (1961).

Aristotle, *Metaphysics*, in Clagett (1961).

Armstrong, D. (1983) *What is a Law of Nature?*, Cambridge: Cambridge University Press.

Arthur, R. (1999) 'On Thought Experiments as A Priori Science', *International Studies in the Philosophy of Science*, vol. 13, no. 3, 215–29.

Aspect, A. *et al.* (1981) 'Experimental Tests of Realistic Theories via Bell's Theorem', *Physical Review Letters*, vol. 47, no. 7, 460–463.

Aspect, A. *et al.* (1982a) 'Experimental Realization of Einstein-Podolsky-Rosen *Gedankenexperiment*: A New Violation of Bell's Inequalities', *Physical Review Letters*, vol. 49, no. 2.

Aspect, A. *et al* (1982b) 'Experimental Test of Bell's Inequalities Using Time-Varying Analyzers', *Physical Review Letters,* vol. 49, no. 25.

Ayer, A.J. (1956) 'What is a Law of Nature?' reprinted in T. Beauchamp (ed.) *Philosophical Problems of Causation,* Encino, CA: Dickenson, 1974.

Bell, J.S. (1987) *Speakable and Unspeakable in Quantum Mechanics,* Cambridge: Cambridge University Press.

Benacerraf, P. (1964) 'What Numbers Could Not Be', in Benacerraf and Putnam (1983).

Benacerraf, P. (1973) 'Mathematical Truth', in Benacerraf and Putnam (1983).

Benacerraf, P. and H. Putnam (eds) (1983) *Philosophy of Mathematics,* Cambridge: Cambridge University Press.

Bishop, M. (1998) 'An Epistemological Role for thought Experiments', in N. Shanks (ed.), *Idealization IX: Idealization in Contemporary Physics*, Amsterdam: Rodopoi, pp. 19–33

Bishop, M. (1999) 'Why Thought Experiments are Not Arguments', *Philosophy of Science*, vol. 66, 534–41.

Bohm, D. (1951) *Quantum Theory,* Englewood Cliffs, NJ: Prentice-Hall.

Bohm, D. (1952) 'A Suggested Interpretation of the Quantum Theory in Terms of "Hidden" Variables, I and II', *Physical Review*, vol. 85, 166–193.

Bohr, N. (1949) 'Discussions with Einstein on Epistemological Problems in Atomic Physics' in A. Schilpp (ed.) *Albert Einstein: Philosopher-Scientist*, La Salle, IL: Open Court.

Bokulich, A. (2001) 'Rethinking Thought Experiments', *Perspectives on Science*, vol. 9, no. 3, 285–307.

Born, M. (1956) *Physics in My Generation*, London: Pergamon Press.

Born, M. (1971) *The Born-Einstein Letters*, New York: Walker and Co.

Braithwaite, R.B. (1953) *Scientific Explanation*, Cambridge: Cambridge University Press.

Brendel, Elke (2004) 'Intuition Pumps and the Proper Use of Thought Experiments', *Dialectica*, vol. 58, no. 1, 88–108.

Brown, James Robert (1986) 'Thought Experiments Since the Scientific Revolution', *International Studies in the Philosophy of Science*, vol. 1, 1–15.

Brown, James Robert (1993) 'Why Empiricism Won't Work',*Proceedings of the Philosophy of Science Association*, vol. 2, 271–9.

Brown, James Robert (2004) 'Why Thought Experiments Transcend Experience', in C. Hitchcock (ed.), *Contemporary Debates in the Philosophy of Science*, Malden, MA: Blackwell, pp. 23–43.

Brown, James Robert (2007a) 'Counter Thought Experiments' (Royal Institute of Philosophy Lecture), *Philosophy* 155–177.

Brown, James Robert (2007b) 'Thought Experiments in Science, Philosophy, and Mathematics', *Croatian Journal of Philosophy*, vol. 7, no. 19, 3–28 [This is part of special issue in two parts, nos. 19 and 20, with my replies.]

Brown, James Robert (2008) *Philosophy of Mathematics: A Contemporary Introduction to the World of Proofs and Pictures*, London and New York: Routledge.

Bunzl, Martin (1996) 'The Logic of Thought Experiments', *Synthese*, vol. 106, no. 2, 227–40.

Buridan, Jean, 'Questions on the Eight Books of the Physics of Aristotle', in Clagett (1961).

Buridan, Jean, 'Questions on Aristotle's *On the Heavens*', in Clagett (1961).

Butts, R.E. and J. Pitt (1978) *New Perspectives on Galileo*, Dordrecht: Reidel.

Buzzoni, Marco (2008) *Thought Experiments in the Natural Sciences: An Operational and Reflexive-Transcendental Conception*, Würzburg: Königshausen & Newmann.

Cargile, James (1987) 'Definitions and Counterexamples', *Philosophy*, vol. 62, 179–93.

Caroti, Stefano (1993) 'Oresme on Motion (Questiones super Physicam, III, 2–7)', *Vivarium*, vol. 31, 8–36.

Chihara, C. (1973) *Ontology and the Vicious Circle Principle*, Ithaca, NY: Cornell University Press.

Chihara, C. (1982) 'A Gödelian Thesis Regarding Mathematical Objects: Do They Exist? And Can We See Them?', *Philosophical Review*, vol. 91, no. 2, 211–27.

Clagett, Marshall (1961) *The Science of Mechanics in the Middle Ages*, Madison: The University of Wisconsin Press.

Cohen, I.B. (1978) (ed.) *Isaac Newton's Papers and Letters on Natural Philosophy*, 2nd edn, Cambridge, MA: Harvard University Press.

Cohen, M. (2005) *Wittgenstein's Beetle and Other Classic Thought Experiments*, Oxford: Blackwell.

Cohnitz, Daniel (2006) *Gedankenexperimente in der Philosophie*, Paderborn: Verlag GmbH.

Cooper, Rachel (2005) 'Thought Experiments', *Metaphilosophy*, vol. 36, no. 3, 328.

Dancy, Jonathan (1985) 'The Role of Imaginary Cases in Ethics', *Pacific Philosophical Quarterly*, vol. 66 (January–April), 141–53.

Davies, D. (2007) 'Thought Experiments and Fictional Narratives', *Croatian Journal of Philosophy*, vol. 7, no. 19, 29–45.

Dennett, D. (1991) *Consciousness Explained*, New York: Little Brown.

Dennett, D. (2005) *Sweet Dreams*, Cambridge, MA: MIT Press.

DePaul, M. and W. Ramsey (eds) (1999) *Rethinking Intuition: The Psychology of Intuition and its Role in Philosophical Inquiry*, New York: Rowman and Littlefield.

Descartes, R. *Principles of Philosophy* (trans. Miller and Miller), *Synthese Library*, Dordrecht: Springer, 1984.

Drake, S. and Drabkin, I.E. (1969) *Mechanics in Sixteenth Century Italy*, Madison: University of Wisconsin Press.

Drake, S. (1978) *Galileo at Work: His Scientific Biography*, Chicago: University of Chicago Press.

Dretske, F. (1977) 'Laws of Nature', *Philosophy of Science*, vol. 44, 248–68.

Dugas, R. (1955) *A History of Mechanics*, Neuchâtel: Editions du Griffon.

Dugas, René (1988) *A History of Mechanics*, New York: Dover.

Duhem, P. (1954) *The Aims and Structure of Physical Theory*, Princeton, NJ: Princeton University Press. (Trans. from the French by P. Weiner.)

Dummett, M. (1967) 'Platonism', in Dummett (1978).

Dummett, M. (1973) 'The Philosophical Basis of Intuitionistic Logic', in Dummett (1978).

Dummett, M. (1978) *Truth and Other Enigmas*, Cambridge, MA: Harvard University Press.

Earman, J. (1986) *A Primer on Determinism*, Dordrecht: Reidel

Eberhard, P. (1977) 'Bell's Theorem Without Hidden Variables', *Il Nuovo Cimento*, vol. 38B, 75–80.

Eberhard, P. (1978) 'Bell's Theorem and the Different Concepts of Locality', *Il Nuovo Cimento*, vol. 46B, 392–419.

Einstein, A. (1905a) 'On the Electrodynamics of Moving Bodies', in Einstein *et al.* (1952).

Einstein, A. (1905b) 'On the Movement of Small Particles Suspended in a Stationary Liquid Demanded by the Molecular-Kinetic Theory of Heat', in Einstein (1956).

Einstein, A. (1905c) 'On a Heuristic Point of View about the Creation and

Conversion of Light', in ter Haar (ed.) *The Old Quantum Theory*, London: Pergamon Press, 1967.

Einstein, A. (1916) 'The Foundation of the General Theory of Relativity', in Einstein *et al.* (1952).

Einstein, A. (1918) 'Principles of Research', in Einstein (1954).

Einstein, A. (1919) 'What is the Theory of Relativity?', in Einstein (1954).

Einstein, A. (1921a) 'Geometry and Experience', in Einstein (1954).

Einstein, A. (1921b) 'On the Theory of Relativity', in Einstein (1954).

Einstein, A. (1929) 'On Scientific Truth', in Einstein (1954).

Einstein, A. (1931) 'Maxwell's Influence on the Evolution of the Idea of Physical Reality', in Einstein (1954).

Einstein, A. (1933) 'On the Method of Theoretical Physics', in Einstein (1954).

Einstein, A. (1934a) 'The Problem of Space, Ether, and the Field in Physics', in Einstein (1954).

Einstein, A. (1934b) 'Notes on the Origin of the General Theory of Relativity', in Einstein (1954).

Einstein, A. (1935) 'Physics and Reality', in Einstein (1954).

Einstein, A. (1940) 'The Fundamentals of Theoretical Physics', in Einstein (1954).

Einstein, A. (1949) 'Autobiographical Notes', in Schilpp (1949).

Einstein, A. (1954) *Ideas and Opinions*, New York: Bonanza Books.

Einstein, A. (1956) *Investigation on the Theory of Brownian Movement*, New York: Dover.

Einstein, A. and L. Infeld (1938) *The Evolution of Physics*, New York: Simon and Schuster.

Einstein, A., B. Podolsky, and N. Rosen (1935) 'Can Quantum Mechanical Description of Reality Be Considered Complete?', *Physical Review*, vol. 47, 777.

Einstein *et al.* (1952) *The Principle of Relativity*, New York: Dover.

Elgin, C. (2007) 'The Laboratory of the Mind,' in J. Gibson, W. Huemer and L. Pocci (eds) *A Sense of the World*, London: Routledge.

Espagnat, B.d' (1979) 'Quantum Mechanics and Reality', *Scientific American*, November.

Espagnat, B.d' (1976) *Conceptual Foundations of Quantum Mechanics*, 2nd edn, London: Benjamin.

Feferman, S. (1983) 'Kurt Gödel: Conviction and Caution', reprinted in S. Shanker (ed.) *Gödel's Theorem in Focus*, London: Croom Helm, 1988.

Feyerabend, P. (1975) *Against Method*, London: New Left Books.

Feyerabend, P. (1980) 'Zahar on Mach, Einstein and Modern Science', *British Journal for the Philosophy of Science*, vol. 31, no. 3, 273–82.

Feyerabend, P. (1984) 'Mach's Theory of Research and its Relation to Einstein', *Studies in the History and Philosophy of Science*, vol. 15, 1–22.

Feynman, R., R. Leighton and M. Sands (1963) *The Feynman Lectures in Physics* (3 vols), Reading, MA: Addison-Wesley.

214   *Bibliography*

Field, H. (1980) *Science Without Numbers*, Oxford: Blackwell.

Fine, A. (1986) *The Shaky Game: Einstein, Realism and the Quantum Theory*, Chicago: University of Chicago Press. Fischbach, E. *et al.* (1986) 'Reanalysis of the Eötvos Experiment', *Physical Review Letters*, Jan.

Frank, P. (1947) *Einstein: His Life and Times*, New York: Knopf.

Friedman, M. (1983) *Foundations of Space-Time Theories*, Princeton, NJ: Princeton University Press.

Galileo, *Dialogue Concerning the Two Chief World Systems* (trans. from the *Dialogo* by S. Drake), 2nd revised edn, Berkeley: University of California Press, 1967.

Galileo, *Two New Sciences* (trans. from the *Discorsi* by S. Drake), Madison: University of Wisconsin Press, 1974.

Geach, P. (1969) *God and the Soul*, London: Routledge and Kegan Paul.

Gendler, T.S. (1998) 'Galileo and the Indispensability of Scientific Thought Experiment', *The British Journal for the Philosophy of Science*, vol. 49, no. 3, 397–424.

Gendler, T.S. (2000) *Thought Experiment: On the . . . Powers and Limits of Imaginary Cases*. New York: Garland Press (now Routledge).

Gendler, T.S. (2002a) 'Personal Identity and Thought-Experiments', *Philosophical Quarterly*, vol. 52, no. 206, 34–54.

Gendler, T.S. (2004) 'Thought Experiments Rethought—and Reperceived', *Philosophy of Science*, vol. 71, 1152–64.

Gendler, T.S. and J. Hawthorne (eds) (2002) *Conceivability and Possibility*. New York, Oxford: Clarendon/Oxford University Press.

Genz, H. (1999) *Gedankenexperimente*, Weinheim: Wiley-VCH.

George, R. (forthcoming) 'Thought Experiments and Arguments in Epistemology', in Rescher (ed.) *Thought Experiments*.

Giaquinto, Marcus (2007) *Visual Thinking in Mathematics*. Oxford: Clarendon Press.

Gödel, K. (1944) 'Russell's Mathematical Logic', reprinted in P. Benacerraf and H. Putnam, *Philosophy of Mathematics*, Cambridge: Cambridge University Press, 1983.

Gödel, K. (1947) 'What is Cantor's Continuum Problem?', in Benacerraf and Putnam (1983).

Goldman, A. (1967) 'A Causal Theory of Knowing', *Journal of Philosophy*, vol. 64, no. 12, 355–72.

Gooding, D. (1992) 'The Cognitive Turn, or, Why Do Thought Experiments Work?' In Giere (ed.) *Cognitive Models of Science*, Minneapolis: University of Minnesota Press, pp.45–76.

Gooding, D. (1993) 'What is Experimental About Thought Experiments?' in D. Hull, M. Forbes, and K. Okruhlik (eds) *PSA 1992*, vol. 2, East Lansing, MI: Philosophy of Science Association, pp. 280–90.

Gooding, D. (1994) 'Imaginary Science', *British Journal for the Philosophy of Science*, vol. 45, no. 4, 1029–45.

Goodman, N. (1947) 'The Problem of Counterfactual Conditionals', reprinted in *Fact, Fiction, and Forecast*, Indianapolis: Bobbs-Merrill, 1965.

Grant, Edward (1971) *Nicole Oresme and the Kinematics of Circular Motion*, Madison: University of Wisconsin Press.

Grant, Edward (1981) *Much Ado About Nothing: Theories of Space and Vacuum From the Middle Ages to the Scientific Revolution*, Cambridge: Cambridge University Press.

Grant, Edward (2002) 'Medieval Natural Philosophy: Empiricism Without Observation', in *The Dynamics of Aristotelian Natural Philosophy from Antiquity to the Seventeenth Century*, vol. 5, edited by C. Leijenhorst *et al.*, Leiden: Brill, pp. 141–68.

Gribanov, D.P. *et al.* (1979) *Einstein and the Philosophical Problems of 20th-Century Physics*, Moscow: Progress.

Grünbaum, A. (1973) *Philosophical Problems of Space and Time*, Dordrecht: Reidel.

Hacking, I. (1983) *Representing and Intervening*, Cambridge: Cambridge University Press.

Hacking, I. (1993) 'Do Thought Experiments have a Life of Their Own?' in D. Hull, M. Forbes, and K. Okruhlik (eds) *PSA 1992*, vol. 2, East Lansing, MI: Philosophy of Science Association, pp. 302–8.

Haggqvist, S. (1996) *Thought Experiments in Philosophy*, Stockholm: Almqvist & Wiksell International.

Hale, B. (1987) *Abstract Objects*, Oxford: Blackwell.

Hardy, G.H. (1929) 'Mathematical Proof', *Mind*, vol. 38, no. 149, 1–25.

Harman, G. (1973) *Thought*, Princeton, NJ: Princeton University Press.

Harper, W. (1989) 'Consilience and Natural Kind Reasoning', in J. Brown and J. Mittelstrass (eds) *An Intimate Relation: Studies in the History and Philosophy of Science Presented to R.E. Butts on his 60th Birthday*, Dordrecht: Reidel.

Harré, R. (1981) *Great Scientific Experiments*, Oxford: Oxford University Press.

Heisenberg, W. (1930) *The Physical Principles of the Quantum Theory*, Chicago: University of Chicago Press.

Heisenberg, W. (1958) *Physics and Philosophy*, New York: Harper and Row.

Heisenberg, W. (1971) *Physics and Beyond*, New York: Harper and Row.

Hempel, C. (1965) 'Typological Methods in the Natural and Social Sciences', in *Aspects of Scientific Explanation*, New York: Free Press.

Hesse, M. (1961) *Forces and Fields*, Totawa, NJ: Littlefield.

Holton, G. (1960) 'On the Origins of the Special Theory of Relativity', in Holton (1973).

Holton, G. (1964) 'Poincaré and Relativity', in Holton (1973).

Holton, G. (1967) 'Influences on Einstein's Early Work', in Holton (1973).

Holton, G. (1968) 'Mach, Einstein, and the Search for Reality', in Holton (1973).

Holton, G. (1969) 'Einstein, Michelson, and the "Crucial Experiment"', in Holton (1973).

Holton, G. (1973) *Thematic Origins of Scientific Thought: Kepler to Einstein*, Cambridge, MA: Harvard University Press.

Holton, G. (1979) 'Einstein's Model for Constructing a Scientific Theory', in Holton (1986).

Holton, G. (1980) 'Einstein's Scientific Programme: The Formative Years', in Holton (1986).

Holton, G. (1981) 'Einstein's Search for a *Weltbild*', in Holton (1986).

Holton, G. (1982) 'Einstein and the Shaping of Our Imagination', in Holton (1986).

Holton, G. (1986) *The Advancement of Science and Its Burdens,* Cambridge: Cambridge University Press.

Holton, G. and Y. Elkana (1982) (eds) *Albert Einstein: Historical and Cultural Perspectives,* Princeton, NJ: Princeton University Press.

Horowitz, T. and G. Massey (eds) (1991) *Thought Experiments in Science and Philosophy*, Savage, MD: Rowman and Littlefield.

Horwich, P. (1982) 'Three Forms of Realism', *Synthese*, vol. 52, 181–201.

Howard, D. (1984) 'Realism and Conventionalism in Einstein's Philosophy of Science: The Einstein-Schlick Correspondence', *Philosophia Naturalis*, vol. 21, 618–29.

Howard, D. (1985) 'Einstein on Locality and Separability', *Studies in the History and Philosophy of Science*, vol. 16, 171–201.

Hume, D. (1975) in P.H. Nidditch (ed.) *Enquiry Concerning Human Understanding,* Oxford University Press.

Humphries, P. (1994) 'Seven Theses on Thought Experiments', in J. Earman *et al.* (eds) *Philosophical Problems of the Internal and External World*, Pittsburgh: University of Pittsburgh Press, pp. 205–27.

Ierodiakonou, K. (2005) 'Ancient Thought Experiments: A First Approach', *Ancient Philosophy*, vol. 25, 125–40.

Irvine, A. (1986) 'Mathematical Truth and Scientific Realism', unpublished PhD thesis, University of Sydney.

Irvine, A. (1988) 'Epistemic Logicism and Russell's Regressive Method', *Philosophical Studies*, vol. 55, 303–27.

Irvine, A. (1991) 'On the Nature of Thought Experiments in Scientific Reasoning', in Horowitz and Massey (1991)

Jackson, F. (1982) 'Epiphenomenal Qualia', *Philosophical Quarterly*, vol. 32, no. 127, 127–36.

Jammer, M. (1974) *The Philosophy of Quantum Mechanics,* New York: Wylie.

Jarrett, J. (1984) 'On the Physical Significance of the Locality Conditions in the Bell Arguments', *Nous*, vol. 18, 569–89.

Johnson-Laird, P. (1983) *Mental Models: Towards a Cognitive Science of Language, Inference, and Consciousness.*

Johnson-Laird, P. (1989) 'Mental models', in Posner (ed.) *Foundations of Cognitive Science.*

Jordan, T.F. (1982) 'Quantum Correlations Do Not Transmit Signals', *Physics Letters,* vol. 94A, 264.

King, Peter (1991) 'Mediaeval Thought-Experiments: The Metamethodology of Mediaeval Science', in *Thought Experiments in Science and Philosophy*, edited by Tamara Horowitz and Gerald J. Massey, pp. 43–64.

Kitcher, P. (1983) *Mathematical Knowledge,* Oxford: Oxford University Press.

Klassen, S. (2006) 'The Science Thought Experiment: How Might it be Used Profitably in the Classroom?' *Interchange*, vol. 37, no. 1, 77–96.

Koyré, A. (1965) *Newtonian Studies*, Chicago: University of Chicago Press.

Koyré, A. (1968) 'Galileo's Treatise "De Motu Gravium": The Use and Abuse of Imaginary Experiment', in *Metaphysics and Measurement,* London: Chapman and Hall.

Kripke, S. (1980) *Naming and Necessity,* Oxford: Blackwell.

Kuhn, T. (1964) 'A Function for Thought Experiments', reprinted in Kuhn, *The Essential Tension,* Chicago: University of Chicago Press, 1977.

Kuhn, T. (1970) *The Structure of Scientific Revolutions,* 2nd edn, Chicago: University of Chicago Press.

Kühne, U. (2005) *Die Methode des Gedankenexperiments*, Frankfurt: Suhrkamp.

Kujundzic, Nebojsa (1992) 'How Does the Laboratory of the Mind Work?' *Dialogue*, vol. 32, no. 3, 573–78.

Kujundzic, Nebojsa (1995) 'Thought Experiments: Architecture and Economy of Thought', *The Journal of the British Society for Phenomenology*, vol. 26, no. 1, 86–93.

Lakatos, I. (1976) *Proofs and Refutations,* Cambridge: Cambridge University Press.

Landau, L. and E. Lifshitz (1975) *The Classical Theory of Fields* (4th edition), Oxford: Pergamon.

Laudan, L. (1968) The *Vis Viva* Controversy, A Post-mortem', *Isis*, vol. 59, 130–43.

Laymon, R. (forthcoming) 'Thought Experiments by Stevin, Mach and Gouy: Thought Experiments as Ideal Limits and as Semantic Domains', in Rescher (ed.) *Thought Experiments.*

Lear, J. (1977) 'Sets and Semantics', *Journal of Philosophy*, vol. 74, 86–102.

Leibniz, G.W. (1686) 'A Brief Demonstration of a Notable Error of Descartes and Others Concerning a Natural Law', trans. in L. Loemker (ed.) *Leibniz: Philosophical Papers and Letters,* Dordrecht: Reidel, 1970.

Lennox, James G. (1991) 'Darwinian Thought Experiments: A Function for Just-So Stories', in Horowitz and Massey (1991), pp. 223–45.

Lewis, D. (1973) *Counterfactuals,* Oxford: Blackwell.

Lewis, D. (1983) 'Postscript to "Mad Pain and Martian Pain"', in his *Philosophical Papers*, vol. 1, New York: Oxford University Press, pp. 130–2.

Lewis, D. (1986) *On the Plurality of Worlds,* Oxford: Blackwell.

Lewis, D. (1988) 'What Experience Teaches', in *Proceedings of the Russellian Society*, Sydney: University of Sydney. Reprinted in *Mind and Cognition*, W. Lycan (ed.), Oxford: Blackwell, 1990, pp. 499–518.

Lichtenberg, Georg Christoph (1983) *Schriften und Briefe: Sudelbücher,*

Fragmente, Fabeln, Verse (Erster Band), edited by Franz H. Mautner, Frankfurt: Insel Verlag.

Lucretius, *De Rerum Natura*, Loeb Library, Cambridge, MA: Harvard University Press.

McAllister, J.W. (1989) 'Truth and Beauty in Scientific Reason' *Synthese*, vol. 78, 25–51.

McAllister, James (1996) 'The Evidential Significance of Thought Experiments in Science', *Studies in History and Philosophy of Science*, vol. 27, no. 2, 233–50.

McAllister, James (2004) 'Thought Experiments and the Belief in Phenomena', *Proceedings of the 2002 Biennial Meeting of the Philosophy of Science Association, Philosophy of Science*, vol. 71, 1164–75.

McAllister, James (2005) 'The Virtual Laboratory: Thought Experiments in Seventeenth-Century Mechanics', in Helmar Schramm, Ludger Schwarte, and Jan Lazardzig (eds) *Collection, Laboratory, Theater: Scenes of Knowledge in the 17th Century*, New York: Walter de Gruyter; pp. 35–56.

Mach, E. (1960) *The Science of Mechanics*, trans J. McCormack, 6th edn, LaSalle, IL: Open Court.

Mach, E. (1976) 'On Thought Experiments', in *Knowledge and Error*, Dordrecht: Reidel.

Maddy, P. (1980) 'Perception and Mathematical Intuition', *Philosophical Review*, vol. 89, no. 2, 163–96.

Maddy, P. (1984) 'Mathematical Epistemology: What is the Question?' *Monist*, vol. 67, no. 1, 45–55.

Maier, Anneliese (1982a) 'Galileo and the Scholastic Theory of Impetus', in *On the Threshold of Exact Science: Selected Writings of Anneliese Maier on Late Medieval Natural Philosophy*, translated and edited by Steven D. Sargent, Philadelphia: University of Pennsylvania Press, pp. 102–23.

Maier, Anneliese (1982b) 'The Significance of the Theory of Impetus for Scholastic Natural Philosophy', in *On the Threshold of Exact Science: Selected Writings of Anneliese Maier on Late Medieval Natural Philosophy*, translated and edited by Steven D. Sargent, Philadelphia: University of Pennsylvania Press, pp. 76–102.

Maxwell, J.C. (1871) *Theory of Heat*, London: Longman.

Maynard Smith, J. (1978) 'The Evolution of Behavior', *Scientific American*, September.

Miller, A.I. (1981) *Albert Einstein's Special Theory of Relativity*, Reading, MA: Addison-Wesley.

Miller, A.I. (1984) *Imagery in Scientific Thought*, Boston: Birkhauser.

Miščević, Nenad (1992) 'Mental Models and Thought Experiments', *International Studies in the Philosophy of Science*, vol. 6, no. 3, 215–26.

Miščević, Nenad (1997) 'Categorial and Essentialist Intuitions: A Naturalist Perspective', *Acta Analytica*, vol. 12, no. 19, 21–39.

Miščević, Nenad (2007) 'Modelling Intuitions and Thought Experiments', *Croatian Journal of Philosophy*, vol. 7, no. 20, 181–214.

Mumford, D. (2000) 'Dawning of the Age of Stochasticity' in V. Arnold (ed.)

Mathematics: Frontiers and Perspectives, New York: American Mathematical Society.

Nersessian, Nancy (1992) 'How Do Scientists Think? Capturing the Dynamics of Conceptual Change in Science', in R. Giere (ed.), *Cognitive Models of Science*, Minneapolis: University of Minnesota Press, pp. 3–44.

Nersessian, Nancy (1993) 'In the Theoretician's Laboratory', in D. Hull, M. Forbes, and K. Okruhlik (eds) *PSA 1992*, vol. 2, East Lansing, MI: Philosophy of Science Association, pp. 291–301.

Newton, I. *Mathematical Principles of Natural Philosophy*, translated from the *Prindpia* (1686) by Mott and Cajori, Berkeley: University of California Press, 1934.

Newton-Smith, W.H. (1982) *The Rationality of Science*, London: Routledge & Kegan Paul.

Norton, J. (1985) 'What was Einstein's Principle of Equivalence?', *Studies in the History and Philosophy of Science*, vol. 16, 203–46.

Norton, J. (1991) 'Thought Experiments in Einstein's Work', in T. Horowitz and G. Massey (eds) *Thought Experiments in Science and Philosophy*, Savage, MD: Rowman and Littlefield.

Norton, J. (1996) 'Are Thought Experiments Just What You Thought?' *Canadian Journal of Philosophy*, vol. 26, 333–66.

Norton, J. (2004a) 'On Thought Experiments: Is There More to the Argument?' *Proceedings of the 2002 Biennial Meeting of the Philosophy of Science Association, Philosophy of Science*, vol. 71, 1139–51.

Norton, J. (2004b) 'Why Thought Experiments Do Not Transcend Empiricism', in Christopher Hitchcock (ed.) *Contemporary Debates in the Philosophy of Science*, Oxford: Blackwell, pp. 44–66.

Oresme, Nicole, *Le Livre du ciel et du monde*, edited by Albert D. Menut and Alexander J. Denomy, Detached from *Mediaeval Studies*, vols 3–5, 1941–1943, New York, 1943.

Oresme, Nicole, *Le Livre du ciel et du monde*, translated and edited by B. Tolley, in *Physical Thought from the Presocratics to the Quantum Physicists: An Anthology*, edited by Shmuel Sambursky, London: Hutchinson, 1974.

Oresme, Nicole, *Questiones super De cello,* translated and edited by B. Tolley, in *Physical Thought from the Presocratics to the Quantum Physicists: An Anthology*, edited by Shmuel Sambursky' London: Hutchinson, 1974.

Pais, A. (1982) *Subtle is the Lord,* Oxford: Oxford University Press.

Papineau, D. (1977) 'The *Vis Viva* Controversy', *Studies in the History and Philosophy of Science*.

Papineau, D. (1987) 'Mathematical Fictionalism', *International Studies in the Philosophy of Science*.

Pappas, G. and M. Swain (eds) (1978) *Essays on Knowledge and Justification,* Ithaca, NY: Cornell University Press.

Parfit, Derek (1984/87) *Reasons and Persons,* Oxford: Clarendon Press.

Parsons, C. (1980) 'Mathematical Intuition', *Proceedings of the Aristotelian Society, 1979–1980.*

Pedersen, Olaf (1993) *Early Physics and Astronomy: A Historical Introduction,* Cambridge: Cambridge University Press.

Peierls, R. (1979) *Surprises in Theoretical Physics,* Princeton, NJ: Princeton University Press.

Peirce, C.S. (1931–35) 'The Reality of Thirdness' in *Collected Papers of Charles Sanders Peirce,* ed. C. Hartshorne and P. Weiss, Cambridge, MA: Harvard University Press, vol. V, 64–67.

Perry, J. (1975) *Personal Identity,* Berkeley: University of California Press.

Poincaré, H. (1952a) 'The Nature of Mathematical Reasoning', in *Science and Hypothesis,* New York: Dover.

Poincaré, H. (1952b) 'Non-Euclidean Geometries', in *Science and Hypothesis,* New York: Dover.

Poincaré, H. (1958) 'Intuition and Logic in Mathematics', in *The Value of Science,* New York: Dover.

Popper, K. (1959) *The Logic of Scientific Discovery,* London: Hutchinson.

Popper, K. (1972) *Objective Knowledge,* Oxford: Oxford University Press.

Popper, K. (1982) *Quantum Theory and the Schism in Physics,* Totawa, NJ: Rowman and Littlefield.

Popper, K. and J. Eccles (1977) *The Self and Its Brain,* London, Berlin, New York: Springer-Verlag.

Prizibram, K. (1967) (ed.) *Letters on Wave Mechanics,* New York: Philosophical Library.

Putnam, H. (1975a) *Mathematics, Matter, and Method: Philosophical Papers,* vol. I, Cambridge: Cambridge University Press.

Putnam, H. (1975b) *Mind, Language and Reality: Philosophical Papers,* vol. II, Cambridge: Cambridge University Press.

Putnam, H. (1975c) 'The Meaning of Meaning', in Putnam *Mind, Language and Reality: Philosophical Papers,* vol. II, Cambridge: Cambridge University Press.

Putnam, H. (1983) *Realism and Reason: Philosophical Papers,* vol. III, Cambridge: Cambridge University Press.

Quine, W. (1960) *Word and Object,* Cambridge, MA: MA Press.

Ramsey, F.P. (1931) 'General Propositions and Causality', in *The Foundations of Mathematics,* London: Routledge & Kegan Paul.

Recami, E. (1986) *Revista del Nuova Cimento,* vol. 9, no. 6.

Redhead, M. (1987) *Incompleteness, Nonlocality, and Realism,* Oxford: Oxford University Press.

Reichenbach, H. (1949) 'The Philosophical Significance of the Theory of Relativity', in A.Schilpp (ed.) *Albert Einstein: Philosopher-Scientist,* La Salle, IL: Open Court.

Reichenbach, H. (1974) *The Philosophy of Space and Time,* trans. from the German original of 1927 by M. Reichenbach and J. Freund, New York: Dover.

Rescher, N. (ed.) (forthcoming) *Thought Experiments.*

Rescher, N. (2005) *What If?: Thought Experimentation in Philosophy,* New Brunswick, NJ: Transaction Publishers.

Rosenthal-Schneider, I. (1980) *Reality and Scientific Truth*, Detroit: Wayne State University Press.

Russell, B. (1897) *The Foundations of Geometry*, Cambridge: Cambridge University Press.

Schildknecht, Christiane (1990) *Philosophische Masken: Literarische Formen der Philosophie bei Platon, Descartes, Wolff und Lichtenberg*, Stuttgart: Metzler.

Schilpp, A. (ed.) (1949) *Albert Einstein: Philosopher-Scientist*, La Salle, IL: Open Court.

Schlick, M. (1920) *Space and Time in Contemporary Physics*, trans. H. Brose, Oxford: Oxford University Press.

Schrödinger, E (1935) 'The Present Situation in Quantum Mechanics', trans. and reprinted in J. Wheeler and W. Zurek, *Quantum Theory and Measurement*, Princeton, NJ: Princeton University Press.

Searle, J. (1980) 'Minds, Brains, and Programs', reprinted in J. Haugeland (ed.) *Mind Design*, Cambridge, MA: MIT Press, 1982.

Shankland, R.S. (1963) 'Conversations with Einstein', *American Journal of Physics*, vol. 31, no. 1, 47–57.

Shankland, R.S. (1973) 'Conversations with Einstein, II', *American Journal of Physics*, vol. 41, 895–901.

Shimony, A. (1978) 'Metaphysical Problems in the Foundations of Quantum Mechanics', *International Philosophical Quarterly*, vol. 18, 3–17.

Shimony, A. (1984) 'Contextual Hidden Variables and Bell's Inequality', *British Journal for the Philosophy of Science*, vol. 35, no. 1, 25–45.

Shimony, A. (1986) 'Events and Processes in the Quantum World', in R. Penrose and C.J. Isham (eds) *Quantum Concepts in Space and Time*, Oxford: Oxford University Press.

Shoemaker, S. (1969) 'Time Without Change', *Journal of Philosophy* vol. 66, 363–81.

Sklar, L. (1976) *Space, Time, and Spacetime*, Berkeley: University of California Press.

Solovine, M. (1987) *Albert Einstein: Letters to Solovine*, New York: Philosophical Books (trans. from the French by W. Baskin).

Sorensen, Roy (1992) *Thought Experiments*. Oxford: Oxford University Press.

Sorensen, Roy (1992b) 'Thought Experiments and the Epistemology of Laws', *Canadian Journal of Philosophy*, vol. 22, no. 1, 15–44.

Steiner, M (1975) *Mathematical Knowledge*, Ithaca, NY: Cornell University Press.

Stinner, A. (1990) 'Philosophy, Thought Experiments, and Large Context Problems in the Secondary Physics Course', *International Journal of Science Education*, vol. 12, no. 3, 244–257.

Strawson, P. (1959) *Individuals: An Essay in Descriptive Metaphysics*, London: Methuen.

Swirski, Peter (2007) *Of Literature and Knowledge: Explorations in Narrative Thought Experiments, Evolution and Game Theory*, London, New York: Routledge.

Taylor, E.F. and J. Wheeler (1963) *Spacetime Physics*, San Francisco: Freeman.

Thijssen, J. (2000) 'Late-Medieval Natural Philosophy: Some Recent Trends in Scholarship', *Recherches de Théologie et Philosophie médiévales*, vol. 67, 158–90.

Thomson, J. (1971) 'A Defence of Abortion', reprinted in J. Feinberg (ed.) *The Problem of Abortion*, 2nd edn, Belmont, CA: Wadsworth, 1984.

Tittle, P. (2005) *What If ... Collected Thought Experiments in Philosophy*, New York: Pearson Longman.

Tooley, M. (1977) 'The of Nature of Laws', *Canadian Journal of Philosophy*, vol. 7, 667–98.

Tooley, M. (1984) 'A Defense of Abortion and Infanticide', in J. Feinberg (ed.) *The Problem of Abortion*, 2nd edn, Belmont, CA: Wadsworth.

Urbaniec, J. (1988) 'In Search of a Philosophical Experiment', *Metaphilosophy*, vol. 19, 294–306.

Urquhart, A. (1989) 'The Logic of Physical Theory', in A. Irvine (ed.) *Physicalism in Mathematics*, Dordrecht: Kluwer.

Van Fraassen, B. (1980) *The Scientific Image*, Oxford: Oxford University Press.

Van Fraassen, B. (1982) 'The Charybdis of Realism: Epistemological Implications of Bell's Inequality', *Synthese*, vol. 52, 25–38.

Wang, H. (1974) *From Mathematics to Philosophy*, London: Routledge & Kegan Paul.

Wang, H. (1987) *Reflections on Kurt Gödel*, Cambridge, MA: MIT Press.

Weisskopf, V.F. (1960) 'The visual appearance of rapidly moving objects', *Physics Today*.

Wheeler, J. and W. Zurek (eds) (1983) *Quantum Theory and Measurement*, Princeton, NJ: Princeton University Press.

Wigner, E. (1962) 'Remarks on the Mind-Body Question', reprinted in *Symmetries and Reflections*, Cambridge, MA: MIT Press, 1970.

Wilkes, K.V. (1988) *Real People: Philosophy of Mind Without Thought Experiments*, Oxford: Oxford University Press.

Williamson, T. (2007) *The Philosophy of Philosophy*, Oxford: Oxford University Press.

Wolters, G. (1984) 'Ernst Mach and the Theory of Relativity', *Philosophia Naturalis*, vol. 21, 630–41.

Wright, C. (1983) *Frege's Conception of Numbers as Objects*, Aberdeen: Aberdeen University Press.

Yablo, Stephen (1993) 'Is Conceivability a Guide to Possibility?' *Philosophy and Phenomenological Research*, vol. 53, no. 1, 1–42.

Zahar, E. (1973) 'Why Did Einstein's Programme Supersede Lorentz's?', *British Journal for the Philosophy of Science*, vol. 24, no. 2, 95–123.

Zahar, E. (1977) 'Mach, Einstein, and the Rise of Modern Science', *British Journal for the Philosophy of Science*, vol. 31, no. 3, 273–82.

Zahar, E. (1980) 'Einstein, Meyerson and the Role of Mathematics in Physical Discovery', *British Journal for the Philosophy of Science*, vol. 31, no. 1, 1–43.

Index

224   *Index*